Best Practice Guide on the Management of Metals in Small Water Supplies

Best Practice Guides on Metals and Related Substances in Drinking Water

Best Practice Guide on the Management of Metals in Small Water Supplies

Edited by
Matthew Bower and Dr Colin Hayes

IWA PUBLISHING

Published by **IWA Publishing**
Alliance House
12 Caxton Street
London SW1H 0QS, UK
Telephone: +44 (0)20 7654 5500
Fax: +44 (0)20 7654 5555
Email: publications@iwap.co.uk
Web: www.iwapublishing.com

First published 2016

British Library Cataloguing in Publication Data
A CIP catalogue record for this book is available from the British Library

ISBN: 9781780406398 (Paperback)
ISBN: 9781780406404 (eBook)

Contents

Chapter 9

Authors

Editors

Matthew Bower, Drinking Water Quality Regulator for Scotland, UK
Colin Hayes, Swansea University, UK

Authors and Contributors

Giovanni Arena, Catania University, Italy
Michela Biasibetti, University of Brescia, Italy
Matthew Bower, Drinking Water Quality Regulator for Scotland, UK
Gea Oliveri Conti, Catania University, Italy
Chiara Copat, Catania University, Italy
Letizia Ferlito, SNV Netherlands Development Agency, Netherlands
Margherita Ferrante, Catania University, Italy
Francesca Gialdini, University of Brescia, Italy
Alfina Grasso, Catania University, Italy
Tom Hall, WRC, UK
Colin Hayes, Swansea University, UK
Jun Ma, Harbin Institute of Technology, China
John Murphy, Springhead Water Consulting, UK
Colette Robertson-Kellie, Drinking Water Quality Regulator for Scotland, UK
Tim Shakesby, Shakesby Pumps, UK
Patricia Sheldon, Highland Council, UK
Sabrina Sorlini, University of Brescia, Italy

Acknowledgements

Thanks are due to all the individuals who contributed to this publication, especially those who offered assistance at relatively short notice. The authors would also like to express their gratitude to the many organisations who have provided access to data and information and allowed these to be used in this publication.

We would also like to acknowledge the support of the IWA Metals and Related Substances Specialist Group, whose management committee assisted with the peer review process. Their constructive comments were gratefully received.

Review Panel

The authors and publishers wish to thank all those involved in the review of this Best Practice Guide. The review involved the following:

Tiina Leiviskä, University of Oulu, Finland
Mike Schock, US Environmental Protection Agency, USA
Tom Sorg, US Environmental Protection Agency, USA
Maronel Steyn, Council for Scientific and Industrial Research (CSIR), South Africa

Abbreviations and Acronyms

CR	Continuous Regeneration
DDS	Domestic Distribution System
DWI	Drinking Water Inspectorate (England and Wales)
DWQR	Drinking Water Quality Regulator (Scotland)
EBCT	Empty Bed Contact Time
EU	European Union
FTU	Formazin Turbidity Unit
GAC	Granular Activated Carbon
ICP	Inductively Coupled Plasma
IR	Intermittent Regeneration
MDPE	Medium Density Polyethylene
NSF	National Science Foundation (USA)
PET	Polyethylene
PoE	Point of Entry
PoU	Point of Use
RO	Reverse Osmosis
SG	Specific Gravity
UV	Ultra-violet
WHO	World Health Organisation
WRAS	Water Regulations Advisory Service (UK)
WSP	Water Safety Plan

About this Best Practice Guide

This Best Practice Guide is one of a series which derives originally from the knowledge assembled by the European research network COST Action 637, supported by a wide range of experts from 26 European countries, the US and Canada. The funding received from COST over the period December 2006 to November 2010 is duly acknowledged. The main objective of COST Action 637 was to stimulate better control of metals in drinking water and to minimise environmental and health impacts. COST is supported by the EU RTD Framework programme and is the oldest and widest European inter-governmental network for co-operation in research.

Since November 2010, the research network has continued to be active as a Specialist Group within the International Water Association, expanding from its primarily European origins to encompass a truly international membership. The aim of the group is to develop and share best practice in the management of metals and related substances in water supplies around the world.

The management of small water supplies presents a unique challenge globally, in countries at all stages of development. A combination of lack of resources, limited understanding of the risks and poor expertise means that individuals and communities may face serious health risks from these supplies. This is not only due to microbiological contamination, but also from contamination by metals, either due to natural or man-made contamination of the source water or through leaching from plumbing materials due to inadequate conditioning and corrosion inhibition and use of inappropriate materials.

This Best Practice Guide aims to share best practice and experience from around the world on a practical level. It will look at general issues relating to small supplies and ways of managing these, adopting a Water Safety Plan approach to deliver sound and lasting improvements to quality. Management techniques and treatment relating to specific metals will be covered, from a theoretical and practical perspective, to deliver a publication that will act as an authoritative guide for all those faced with the problem of ensuring the quality of a small water supply. Varied case-studies will help to illustrate issues and ways in which they have been resolved.

Foreword

Guarding the quality of small water supplies, which are frequently limited in both financial and technical resources, remains a significant challenge for most parts of the world. Although such supplies individually provide water for small numbers of people, the total number supplied is significant and represents the potential for many individuals to be affected by unwanted contaminants. While waterborne pathogens present the greatest acute risk to health, there are chemicals that can have a significant impact on health with repeated exposure.

Metals can reach drinking water from contamination of raw water or by leaching from pipes and fittings. The most widely recognised are arsenic and lead but there are other metals that can either be a threat to health or render water unacceptable for domestic use by affecting colour and, sometimes, taste. It is, therefore, important that there are frameworks for identifying and managing the risks that these contaminants present. This volume provides a timely examination of both the risks and the management approaches that are suitable for application to small supplies in many different settings.

Professor John Fawell
Cranfield University, Water Science Institute.

Executive Summary

The focus of this Best Practice Guide is the management of metals and related substances in small water supply systems. There are many definitions of the term "small water supply" but the majority serve less than 50 people. A common characteristic is weak management, lack of expertise and lack of resources. Governments and regulators around the world have become increasingly aware of the magnitude of the problems associated with small supplies; even so, a large proportion are not subject to regulation despite small water supplies being more likely to contain high concentrations of metals and metalloids, particularly arsenic, copper, iron, lead and manganese.

A drinking water safety plan is a useful tool for assessing and managing the risks associated with small water supplies. Extensive guidance on water safety planning is available from the World Health Organization and the best results come from involving a team of people with an interest in the supply. It will be beneficial for local authorities (or equivalent organisations) to coordinate and prompt water safety planning in their area, thereby making expertise available to individual supply owners. Educational processes have a major part to play in raising the awareness of small supply owners.

Some metals are of health concern whilst others cause nuisance. Metal solubility is strongly influenced by oxidative state, pH, alkalinity and natural organic matter and these factors should feature in risk assessment. Sampling methods have limitations and must be selected with care. Field testing methods are now available and can avoid the high costs associated with laboratory analysis provided they are used correctly.

Many metals in small supplies are present in the source, either naturally occurring or arising from man-made activities. Careful choice of source location and construction of infrastructure can prevent or reduce contamination. Treatment processes can introduce metals into water if inappropriate or poorly controlled. Distribution systems and plumbing materials can also contaminate a supply with metals, notably copper, iron, lead, nickel and zinc. Water conditioning, good system design and careful choice of plumbing materials are key to reducing post-treatment contamination.

The choice of treatment process is usually a compromise between performance, practicalities and cost. Point of entry and point of use systems are often used to treat small water supplies. The appropriate treatment arrangement will depend on individual circumstances. Loose media filters are often used to remove insoluble metals.

Iron is the main metal to affect small water supplies at concentrations requiring treatment. Care should be taken to ensure that oxidation of ferrous iron is substantially complete prior to filtration. Manganese removal also relies on oxidation but can be aided with catalytic removal using manganese dioxide. Arsenic and some other metals can be removed by adsorption onto either activated alumina or ferric oxide. The ongoing maintenance of any installed treatment process is vital in ensuring its continued effective operation.

To minimise metal contamination from distribution pipework and plumbing, it is important to manage the corrosivity of the water, that is, the tendency of the water to dissolve metals. The main approach will be to keep the water's pH above 7.0 (and preferably in the range 7.5 to 8.0) using simple filters containing alkaline media. Additionally, any lead pipes should be replaced using copper or plastic.

For easy reference, a Manual of Individual Metals summarises occurrence, chemistry, health and aesthetic significance, regulatory standards and guideline values, and options for management in a catchment.

Four case studies illustrate how metals in small supplies have been tackled:

- Arsenic removal in Italy
- Iron removal in the United Kingdom
- Metals in areas of water scarcity in Africa
- Unexplained Lead Contamination of a Small Water Supply in Northern Scotland

Chapter 1

The difficulties of managing water quality in small water supplies

1.1 WHAT ARE SMALL WATER SUPPLIES?

1.1.1 Definitions

Size is relative. This applies to most things, and water supplies are no exception. To a municipal water supplier in a large urban area, "small" may mean a water treatment plant less than 1 Ml/d. In many rural areas across the world, there are countless supplies that provide life-giving water to an individual dwelling or a small, remote community. Clearly, a small supply means different things to different people. The USEPA defines a small water supply as serving 501–3300 people, with a supply serving 500 people or less classified as very small. In other parts of the world, a water supply of 500 people would be classed as medium sized or even large. The European Drinking Water Directive legislates for supplies serving more than 50 people or supplying more than 10 cubic metres per day, unless they are involved in a commercial activity.

Such water supplies are often underfunded, poorly managed and neglected, and yet this should not be the case. Even in the developed countries of Western Europe and North America, small water supplies provide the only source of drinking water for a significant minority of the population. A figure of 10% has been estimated for the percentage of the population of the European Union reliant on small or very small water supplies (Hulsman, 2005), based on aggregated data from member states. Globally, the World Health Organisation (WHO) estimates that 884 million people (13% of the world's population) had no access to an improved water supply in 2008 (WHO/UNICEF, 2010). By far the majority of these supplies will be small ones, and most – WHO estimate 94% of this population – will live in rural areas. From similar data on small water supplies around the world it is evident that:

- estimates of the populations served carry a high degree of uncertainty;
- figures vary widely between countries;
- the issue of small, unimproved water supplies is especially acute in rural areas;
- the data on the nature and quality of these supplies is often very limited.

Clearly, in this topic area it is not helpful to rely too closely on definitions and absolute statistics. It is more useful to consider the characteristics of these supplies and the common factors which make the quality of water they supply difficult to maintain consistently to a sufficiently high standard. For the purposes of this publication, small water supplies will be considered primarily to be ones serving less than 500 people and which experience one or more of the following issues:

- minimal or limited regulation
- varying raw water in terms of quantity and quality
- very limited resources for treatment and correct distribution and storage of water
- a lack of expertise in water treatment and supply
- a low awareness of risk
- a lack of robust monitoring and safeguards

With the very smallest water supplies – those serving just a few people – the problems above are often intensified and treatment options more limited. This guide will aim to specifically address the difficulties encountered by these supplies.

1.1.2 What does a small water supply look like?

Sources of raw water for consumption are varied, and most are scaleable, so that a river abstraction could supply a single property or a small town. Similarly, groundwater sources can supply large populations via wellfields containing many boreholes or consist of a single well in someone's garden. Typical sources of water for small supplies include the following:

- Streams
- Lakes or ponds
- Springs
- Wells
- Boreholes
- Temporary sources such as bowser or tanker

There are also a number of management scenarios for small supplies and, more than anything else, it is probably these that define the health risk presented by the supply. A supply serving a handful of people that is owned, managed and operated by a large municipality or utility company is ultimately far less likely to experience the same quality issues at the point of use as those encountered by a supply of the same size that is managed by individuals or a community. This publication is very much aimed at the latter scenario, although the technical challenges described may also affect the former.

Some examples of relevant management scenarios are described below:

Individual responsibility supplies

These supplies are very small and supply a single property or family dwelling. The user of the supply has sole responsibility for the operation and maintenance of the supply, whether the implications of this are fully understood and accepted or not. These supplies are often described in legislation as private wells and may be subject to minimal or no regulation and state or municipal involvement. Water quality monitoring will often be at the discretion of the user, and where there is a lack of interest, awareness of the risks or resources it is highly likely that no monitoring will take place. Where it does take place, the dataset collected is unlikely to be of a sufficient size to provide a representative picture of the quality of the water supplied. Although outside the scope of this publication, the microbiological situation illustrates this point effectively – a single sample taken on one day that is free from faecal indicator bacteria provides no guarantee that the supply is free from contamination on the remaining 364 days of the year. Provided the water appears clean, the assumption may be made that it is safe for drinking.

Figure 1.1 Two examples of a small water supply – one an untreated borehole source in Malawi (left), the other a supply in Scotland with multi-stage filtration, serving an industrial establishment. (*Source*: Drinking Water Quality Regulator for Scotland, 2015.)

Very often these supplies are co-located with other forms of social deprivation and do not feature in nationally compiled data on water quality. This may mean that there is very limited political interest, and consequently little leverage or financial resource to undertake improvements.

It is almost certain that the risk to health presented by these supplies is largely un-quantified and under reported. Attempts to undertake epidemiological studies to link these water supplies to cases of illness are often further hindered by an under-reporting of cases and the presence of other potential environmental sources of disease, such as proximity to livestock.

Community supplies

These supplies cover a range of situations, and go by a variety of names, but all involve a scenario where a group of people become collectively involved in managing a shared water supply. This can result in a range of outcomes, which are usually better than for supplies that are the responsibility of individuals. Reasons for this include:

- Shared resources enable greater purchasing power for treatment and maintenance;
- Such supplies are more likely to have some form of regulation or support available to drive improvement;
- A larger pool of people involved in the supply spreads the workload and increases the likelihood of somebody with an interest in managing the supply

However, such supplies can also experience problems where conflicting interests arise and individuals disagree over approaches or finance. Disputes over water can become very bitter and are often complex to resolve legally.

1.2 THE MANAGEMENT AND REGULATION OF SMALL WATER SUPPLIES

Many governments and global bodies are becoming increasingly aware that a significant proportion of the world's population are reliant on small water supplies, and at an increased health risk. Even in developed countries there is often an "underclass" of water supplies that slips beneath the regulatory "radar". In some countries, Finland for example, regulatory focus has hitherto been on controlling wastewater discharges from single dwellings in remote areas.

Approaches to regulation vary around the world, and some countries are beginning to develop policy and regulatory approaches that cover the smallest supplies. Changing regulations can bring their own problems, with small water supply owners who have invested in improvements early forced to undertake additional work to keep pace. Where the regulatory environment is perceived to be regularly shifting, this can prove to be a dis-incentive to make improvements.

The regulation of small supplies is not easy – data is sparse, making it difficult to administer regulations and monitor their effectiveness. Owners and users of small water supplies may be resistant to regulation – some may be unaware or refuse to accept the risks and see regulatory intervention as an infringement of their personal rights. Successful regulatory approaches around the world will vary depending on cultural and other variables, but the most effective methods are likely to include elements of risk communication and support as well as more traditional numerical standard-based regulation. Many regulators, acknowledging that it is not always feasible, or desirable, to monitor small water supplies to the extent required to assure consistent compliance or safety are advocating a risk-based approach to management, often using the World Health Organisation's Drinking Water Safety Plan (DWSP) methodology. A practical framework for using creating a DWSP for a small water supply system is described in Chapter 2 of this book.

THE REGULATION OF SMALL WATER SUPPLIES IN THE EUROPEAN UNION

The Drinking Water Directive applies to water supplies greater than 50 m^3 per day or serving more than 10 people or supplying a commercial activity. Supplies that fall outside the Directive may be covered by national legislation within individual member states.

Approaches to very small supplies vary across Europe, with many member states making a distinction between small utility/municipal supplies and private wells. Most member states do not have an accurate picture of the number and location of these (Hulsman, 2005), and consequently no concept of water quality or the risk presented by the very smallest supplies.

THE REGULATION OF SMALL WATER SUPPLIES IN THE USA

USEPA regulate public and private water supplies serving greater than 25 people. Special consideration is given to small water supplies in three bands, based on population. When developing new quality standards, the EPA must also consider treatability and the available of appropriate treatment technologies for small water supplies.

Private wells are not regulated by USEPA. Individual states provide guidance and advice, and some states, such as Virginia, have state legislation regulating these supplies.

1.3 THE VULNERABILITY OF SMALL WATER SUPPLIES TO CONTAMINATION BY METALS

The chemical properties and health significance of metals in small water supplies are diverse, and in a similar way the routes by which metals may enter small water supply systems are many and varied.

Superficially, it may appear that contamination routes are the same as for larger supplies, and to some extent this is true, however there are a number of factors that make small supplies particularly vulnerable:

- Small raw water sources will tend to be poorer or more variable in quality
- Greater potential for inappropriate choice of source or abstraction point
- Lack of awareness of risks by owners and users
- Lack of resources and technical knowledge
- Reduced regulation of small supplies
- Reduced frequency and scope of quality monitoring
- Lack of an adequate maintenance programme

The three charts that follow illustrate this point clearly. They show data for iron, manganese and lead from private water supply sampling in one year in Scotland, UK (Drinking Water Quality Regulator for Scotland, 2014). These non-municipal supplies range in size from 0.2 m³/day to 10 m³/day. All would be considered small by most definitions, yet even within this range there is a clear trend of deteriorating quality with decreasing size. This is most clearly seen with manganese, however all datasets exhibited a high degree of variability. The manganese is naturally occurring, while the majority of the lead present in samples is due to the dissolution of lead by low pH, low alkalinity waters with little in the way of conditioning. Iron concentrations may either be due to naturally occurring iron, iron arising from corroding ferrous pipework or a combination of the two. Analysis for all three metals is vulnerable to skewing by the presence of particulate material in the sample, and this may be responsible for the maximum values seen.

For comparison, data for the public water supply in Scotland, managed by Scottish Water, showed quality for the same three parameters to be significantly higher during the same year – the mean concentration for iron, manganese and lead in microgrammes per litre was 15.91, 2.89 and 0.79 respectively.

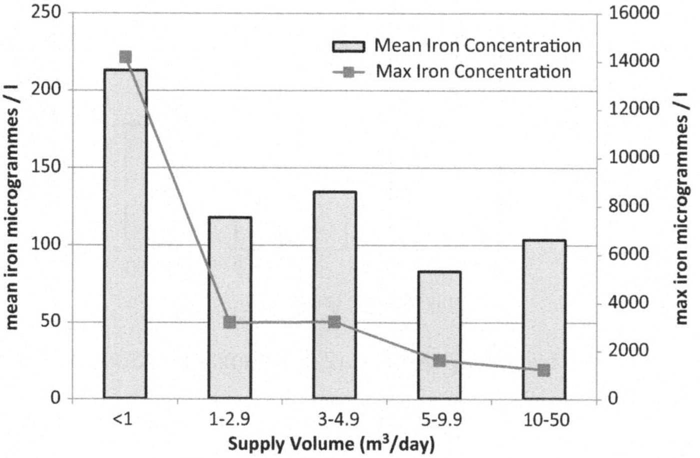

Figure 1.2 Iron concentration by supply volume (1884 samples). (*Source*: Drinking Water Quality Regulator for Scotland, 2015.)

Chapter 4 explores the routes and mechanisms by which metals enter small water supplies. Subsequent chapters consider methods of controlling these mechanisms to prevent them entering the water in the first place and ways by which metals can be removed using an appropriate treatment process. It is therefore only appropriate here to discuss the ways in which these may differ in small water supplies compared with their larger counterparts.

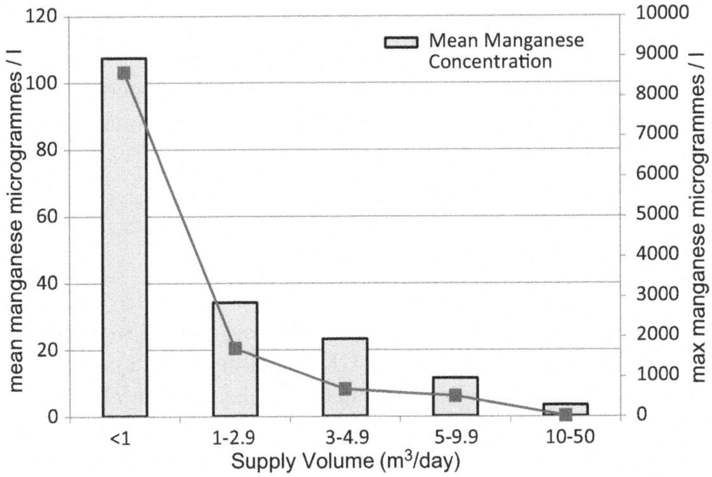

Figure 1.3 Manganese concentration by supply volume (1721 samples). (*Source*: Drinking Water Quality Regulator for Scotland, 2015.)

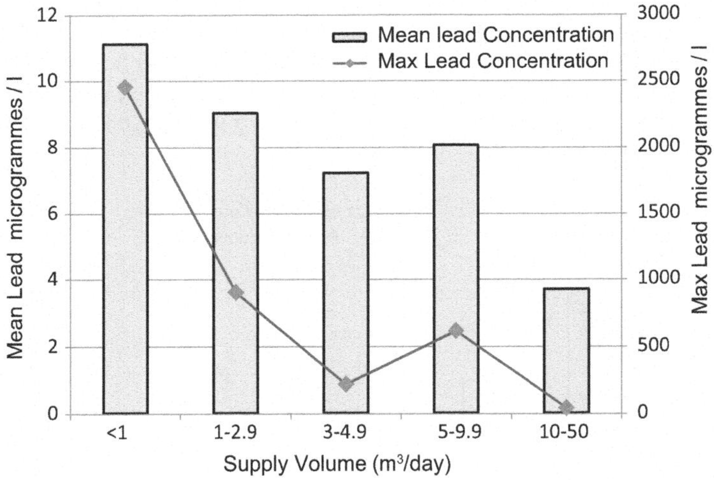

Figure 1.4 Lead concentration by supply volume (2288 samples). (*Source*: Drinking Water Quality Regulator for Scotland, 2015.)

Raw water sources

Small water supplies often make use of the nearest natural watersource – the reason for the choice of source is often lost in the mists of time, and rarely questioned unless someone with some expertise and interest in the supply questions it. This, of course, assumes there is a choice, and in many water scarce areas of the world there is none. Consequently, many small supplies abstract water from inappropriate sources and with little infrastructure to maximise the quality as far as possible. In terms of metals, this may mean water high in iron, manganese and aluminium which cause aesthetic issues, and other substances with implications for health such as arsenic. Industrial pollution, which may not have been present when the supply was constructed, may have encroached on the supply, with heavy metals and other contamination.

Where source waters are concerned, lower volumes often mean poorer quality due to minimal dilution of contaminants. Even where any metals are naturally occurring, small bodies of water are more vulnerable to fluctuations in quality caused due to weather events or seasonal effects such as stratification.

Treatment

Treatment options for small water supplies are often different to those that would be employed on larger, municipal treatment works. Sometimes the larger treatment process isn't scaleable, or it involves dosing of hazardous chemicals to a level of precision that is beyond most people responsible for a small water supply. In developing countries, where many people are reliant on small water supplies, appropriate technology simply isn't available or impractical on cost grounds.

The most successful technologies used on small supplies tend to have certain characteristics in common – they operate reliably, on waters of varying quality with the minimum of monitoring and maintenance. Unfortunately, few technologies exist that meet all of these criteria, and for the removal of some metals, treatment options are limited. The location of many supplies in remote rural areas means that reliable power may be an issue, and this can further limit options. Often the choice of treatment can be a compromise, and it is almost always better to take measures to ensure that the best possible quality source is selected and the way in which water is collected at source further optimises quality. In some cases it can be better, and more cost effective, to abandon a poor quality source rather than pursuing a potentially endless quest to install additional or alternative treatment stages.

Delivery to point of consumption

Clearly there is little point in treating water to a high standard and removing contaminant metals if the storage and pipework post-treatment only serves to re-contaminate the supply or introduce an altogether different set of metals dissolved from the plumbing system. The deficiencies of knowledge, interest and resources that so often hinder the proper treatment of small water supplies also affect the infrastructure post-treatment. Problems can become particularly acute where inadequate treatment causes a lack of conditioning of the water so that it is highly aggressive to plumbing materials. Any water safety plan developed for a supply that does not also consider quality and risks in the distribution pipework and plumbing is only half complete.

KEY POINTS

1. There are many definitions of what constitutes a small water supply.
2. A common characteristic of many small supplies is weak management.
3. Governments and regulators around the world are becoming increasingly aware of the importance and risk presented by small water supplies.
4. Small water supplies are more likely to contain high concentrations of metals.

REFERENCES

Drinking Water Quality Regulator for Scotland (2015). Drinking Water Quality in Scotland 2014 (Annual Report). Drinking Water Quality Regulator for Scotland (DWQR). Edinburgh, Scotland.

Hulsmann A. (2005). Small systems large problems: A European inventory of small water systems and associated problems. Web-based European Knowledge Network on Water (WEKNOW).

US Environmental Protection Agency (2015). Regulating Public Water Systems and Contaminants Under the Safe Drinking Water Act. http://water.epa.gov/lawsregs/rulesregs/regulatingcontaminants/basicinformation.cfm#Are%20there%20special%20considerations%20for%20small%20systems (accessed 15 May 2015).

WHO and UNICEF (2010). Progress on Sanitation and Drinking Water: 2010 Update. World Health Organization & UNICEF, Geneva, New York.

Chapter 2

Water safety plans for small water supplies

2.1 INTRODUCTION

The traditional approach to ensuring drinking water safety was to sample supplies, analyse the samples and then respond to failures of standards. However, by the time the sample had been transported to the laboratory, the analysis completed and the results relayed to the appropriate people, the water had been already been consumed. Only a very small portion of the supply was actually sampled, and the result shows a snapshot of water quality for the moment that the sample has been taken.

The World Health Organisation (WHO) launched guidance on Water Safety Plans (WSP) in their Guidelines for Drinking-water Quality 3rd Edition, and stated '*the most effective means of consistently ensuring the safety of a drinking water supply is through the use of a comprehensive risk assessment and risk management approach that encompasses all steps in water supply from catchment to consumer*' (WHO, 2004).

A WSP is a risk assessment and risk management process for drinking water supplies. Its purpose is to identify risks and ways of managing risk in a co-ordinated and documented manner. A WSP deals with the entire drinking water system, however simple or complex, from the water supply catchment and source, through any treatment, networks and storage to the consumer's tap. It can help ensure that those responsible for or reliant on a drinking water supply understand and manage it so that it is consistently safe for all who use it.

WSP are used by a significant number of water utilities across the world, with many countries enshrining the requirement for WSP in legislation. WHO has published a number of guidance documents on WSP for water utilities. WHO has also widely promoted the use of WSP for small supplies; WHO has long recognised that small drinking water supplies can be a risk to health and has reported '*Community water supplies in both developing and developed countries are more frequently associated with outbreaks of waterborne disease than urban supplies. Investing in these supplies will reduce waterborne disease outbreaks and overall costs.*'.

There is no reason why WSP cannot be developed for small supplies, and their development and maintenance can be critical to preventing waterborne disease from drinking water supplies which are often known to be vulnerable. It is widely recognised that small supplies have challenges that are often not seen with utility company managed supplies; they are less likely to be managed and understood, and there are likely to be limited finances and skilled resources available. However, where there is a will to improve a small supply, to make sustainable, managed and prioritised improvements, there is no doubt that the WSP is the most effective method of achieving the on-going management and safety of small supplies. The principles of preparing and developing WSP are straightforward, and this chapter proposes a simple method of comprehensively capturing and maintaining information on WSP that can be used on any drinking water supply, irrespective of its size or complexity, which is not dependent on the availability of funding, but rather relies on the involvement of the appropriate people and an acknowledgement that drinking water safety is critical to the health and development of communities.

Development of a WSP does not need to be complex or costly, and can, indeed should, develop over time to ensure that it becomes a well understood and trusted management tool that all parties with an interest in the drinking water supply can feel comfortable with. It is inevitable that when a drinking water supply which has not been thoroughly managed is put through the WSP process, the need for investment in the supply may be highlighted. It should also however highlight operational and maintenance requirements, and crucially will help prioritise any investment needs and operational tasks.

This chapter deals with the development and maintenance of a water safety plan for a small water supply. It applies to all risks that may exist on a supply. Chapter 3 provides specific information on the assessment of risks relating to metals.

2.2 MAKING DWSPS WORK FOR SMALL WATER SUPPLIES

It needs to be acknowledged that, in order to be a successful means of safeguarding water quality, a WSP must be accepted and owned by those most closely connected with the water supply. A WSP methodology that is onerous or overly bureaucratic is likely to be doomed to failure, especially where there are minimal drivers in place to encourage the person responsible for the supply to be actively involved in the WSP's production. This is where the regulating or supervising authority has a key role, whether that is a local authority, health authority or national or regional water regulator. Drivers that could be used, either separately or in combination include:

- *Compulsion* – Making the production of a WSP a legal requirement
- *Financial Incentives* – Payment of a grant or provision of other support is conditional upon completion of a satisfactory WSP
- *Education* – Adoption of the WSP approach is far more likely to be successful if people are aware of, and accept, the risks from poor water quality
- *Facilitation* – A simple to use and understand framework for the production of a WSP has a better chance of adoption and ongoing use

In encouraging the use of WSPs, the regulating authority must select the most appropriate drivers based on local cultural, economic, political and regulatory circumstances. Chances of success will be maximised where these are considered at a local level and communities perceive that approaches are not being imposed upon them. Engagement between regulatory bodies and community organisations can often be a useful means of achieving local buy-in. Local knowledge in tandem with accessible expertise can be a powerful combination when it comes to getting things done.

A tiered approach to WSPs is advocated, depending upon the size and complexity of the supply. An overly complex and lengthy plan for a relatively simple supply makes it less likely that the plan will be completed and updated and serve to distance the plan from the non-technical user. Furthermore, significant risks in the supply may be obscured within the detail. It may be feasible to effectively capture, or at least summarise, the main risks and controlling measures for a straightforward water supply on a single page. It is not, however, necessarily safe to assume that the number of water quality risks associated with a supply decrease in proportion to its size, and for some small supplies a quite extensive WSP may be appropriate.

2.3 TEAMWORK – THE VALUE OF A WSP TEAM

It is unlikely that any one person will be capable of developing a WSP on their own. Generally a team is needed – suggested participants could include:

- Those who know the sources of the water supply and the land where the water is collected, for example the land owners, farmers, farm workers. Catchments and sources may be on land which is not owned by the users of the supplies, who in turn may not be aware of risks to the supply associated with their activities.
- Owners of water tanks and pipelines; these may supply more than one property, and individuals may not be aware of tank or pipe materials such as lead which may cause harm to health.
- If there is water treatment on the supply, the individuals who operate and maintain the treatment.
- Those who know where pipes and tanks are – from the source through to any treated water storage tanks, into properties and within buildings. It may not always be apparent who is connected to the supply, which can significantly slow down communication of problems.
- Users of the supply; some may have water treatment, whereas others may not, and they may not be aware of the need for treatment. Some may not know where their water supply originates and where responsibilities lie.
- Individuals with knowledge of drinking water systems, legislative requirements or regulatory responsibility. For example this could be the local authority or health department. They should also be able to interpret analytical and scientific data and information.
- Organisations with environmental data and information, such as environmental regulators; information on the geology of the area may assist in determining potential contaminants to the drinking water supply such as arsenic, iron and manganese. They may also be able to provide information on safer alternative sources or abstraction information.

The best results are achieved by the team working together, getting all points of view across, hearing the issues and needs of all concerned. It is advisable to have nominated individuals who are responsible for the administration tasks, and for setting up and maintaining information and tasks, and it may be beneficial to have a named individual who is responsible for the WSP.

2.4 A PRACTICAL GUIDE TO DEVELOPING A WSP FOR A SMALL SUPPLY

There are many and varied methods for preparing a WSP. They can be electronic, or they can be paper based. It is recommended that a form of version control is adopted, which can be as simple as recording on the WSP when updates are made.

The format of a WSP for a small supply can consist of five sections;

Section	Content
Section 1	Overview of the supply
Section 2	• Risk assessment of the supply from source to tap • Identification of controls for the risks • Methods of monitoring the controls
Section 3	Verification of safety through water sampling and analysis
Section 4	Action Plan to list and prioritise tasks and responsibilities
Section 5	Plan for review of the WSP

Details of each of the content of these sections is given below:

2.4.1 Section 1 – Overview of the supply

Wherever possible, visit the source of the drinking water supply and the catchment and carry out an inspection. The following information should be recorded:

- Documentation of the water supply is crucial, and the creation of procedures to document tasks, both routine and emergency, are very important so that the operation of the supply system is carried out in a consistent and correct manner and can be operated when normal operators are unavailable.
- A map or diagram of the system to record the extent of catchments, the locations of sources and any assets such as tanks and pipework. This could be hand or computer drawn, simple or sophisticated, whatever the team finds most appropriate. It could be written on top of existing published maps or drawn onto on-line satellite images. It should be easy to update as information becomes available; for example if new pipes are laid or tanks are discovered, the plan should be updated. Recording asset locations is very important so that others are able to find and manage them.
- Photographs of the supply system.
- Contact details for individuals or organisations who have responsibilities, such as land owners or anyone involved in the management of the supply or its constituent parts.
- Addresses of all properties served by the supply and contact details of owners. This is essential in the event of there being a water quality problem so that all consumers can be notified. Also if charges are made for the supply of water, all consumers can be fairly billed.
- Names and contact details of manufacturers and contractors for replacement or repair of equipment on the supply, such as pumps and treatment systems, and chemical suppliers, for example for suitable drinking water disinfectants.
- Names and contact details of who can help in emergencies, for example bottled water suppliers, contractors for pipe burst repairs, Environmental Health or health professional.

2.4.2 Section 2 – Risk assessment and identification of controls for risk

The supply should be visited by the team, who should walk the catchment where possible and look at treatment, storage and discuss any domestic distribution before carrying out the risk assessments (Figure 2.1). It may be that not all risks are obvious, and over time, as more knowledge is gathered on the supply, or if problems become apparent, the risk assessment must be updated, and a method for tracking the updates should be devised. This could be as simple or as complex as those managing the WSP feel is appropriate – it could be as simple as putting dates next to newly found risks. Specific information on assessing the risks from metals is provided in Chapter 3.

While all water supplies are different, many of the potential hazards and risks are common to many supplies. Of great use is a template, which can be used to raise pertinent questions about the supply, and to record responses and store information on how risks can be managed. An example of a WSP template is shown in Table 2.2.

Once risks are identified, it is important that methods for managing or controlling risk are identified. Controls can vary widely – they may do not necessarily involve building or installing equipment – often they can be checks or tasks that need to be performed. Also, once the control has been identified, a method of monitoring that control should be determined. Examples are shown on Table 2.2.

Figure 2.1 Visiting the supply and discussing issues is important.

Any available water quality data should be assessed, and the reasons for any water quality failures should be considered, along with consideration of the risk of the issues recurring. Using the scoring system suggested in this chapter, it is likely that any scores over 16 would be copied to the Action Plan (see Section 4) and would need attention. It can be daunting to see the number of risks that are highlighted during a risk assessment, so it is necessary to prioritise risks according to the likelihood and severity of consequences of the risk. Table 2.1 gives an example matrix that can be used by the team to score, and rank risks. This matrix forms part of the Scottish Government's Technical Manual for Private Water Supplies. The use of a scoring system allows risks to be ranked in order of priority, and for resources to be planned.

Table 2.1 Risk matrix.

Likelihood	Severity of Consequences				
	Insignificant	Minor	Moderate	Major	Catastrophic
Almost certain	16	32	64	128	256
Likely	8	16	32	64	128
Moderately likely	4	8	16	32	64
Unlikely	2	4	8	16	32
Rare	1	2	4	8	16

Depending on the use of the supply and its location, for security or confidentiality reasons it may be appropriate to allow carefully considered access to risk assessment information and to protect the information as appropriate.

Table 2.2 Example WSP template.

Risk	Hazard	Control	Control Monitoring	Likelihood	Consequence	Hazard Assessment Score	Comments
Ingress of surface water	Microbiological contamination	Bore well house higher than surrounding ground level and at top of slope, in locked dry building	Daily check on security and dryness of building by J. Masson, signed in WSP diary.	Rare	Major (no disinfection)	8	Historical evidence shows extremely unlikely
Livestock feeding troughs without adequate backflow prevention	Microbiological contamination	Backflow prevention fitted on branches to cattle troughs	Annual inspection of water troughs by L. MacLeod. In WSP diary, signed when completed	Rare	Major	8	Only feeding troughs are at south east of farm – labelled on WSP diagram
Bursts caused by age related deterioration	Loss of supply	Cast iron pipes are thick walled, date from 1940. No known bursts in 20 years	Visual check of site – all users encouraged by annual letter with water bill to contact L. MacLeod if signs of bursts evident. Discuss possibility of replacement of network at next annual meeting (on agenda)	Moderately likely	Major	32	List of contractors to make repairs and bottled water suppliers' details in emergency plan section of WSP
Pipework within farm house lead	Lead	Advised to drink bottled water until resolved. –Plumbers to replace lead pipework to kitchen tap first week of June, upstairs year after due to cost	None. Advice given by health officer for bottled water	Almost certain	Major	128	Water is low pH and is soft. Samples taken failed the standard. Children in household

2.4.3 Section 3 – Verification

The WSP approach has risk assessment and control of risks and hazards at its heart, and its aim is to prevent incidents. However, sampling and analysis remain a vital part of the WSP process, to give verification of water safety. It is a final check and gives reassurance that the supply is safe at the time of sampling. It is recognised that for small rural supplies, the logistics of sampling, the costs of sampling and analysis and the availability of suitable analysing laboratories can be issues, so careful planning of frequency of sampling and parameters to be analysed is essential. Some countries have regulatory requirements for sampling and analysis of small supplies, but others do not, and for very small supplies it may be that there is no sampling at all. The development of a WSP, particularly given the team approach, can be used to determine the most appropriate organisations to take samples, analytical methods, advise on the results of analysis and on any actions that should be taken.

The WSP should identify hazards for individual supplies, but it may be that hazards are not always immediately identified, that they are only picked up through analysis of the supply. Many contaminants cannot be seen or tasted, and may be difficult to predict through risk assessments, particularly if the extent and nature of the catchment is not fully understood, and some contaminants, such as surface water manganese contamination, can be seasonal in nature. If resources allow, it is recommended that a wide range of parameters are considered, and parameters that are not detected can be removed from future sampling schedules.

Sampling of supplies can be done in response to a number of situations. As well as verifying water safety, it can be done for operational reasons, such as confirming the effectiveness of disinfection through testing for chlorine and analysing for microbiology contamination. It can also be done in response to comments from consumers, for example if there is a change in taste, odour or appearance of the drinking water.

For many small water supplies, the costs of sampling and analysis may be a significant issue. Ways of cost-effectively verifying water safety may be explored depending on the situation and any external requirements. Sampling regimes should be carefully considered and planned so that every sample yields representative and useful information about the supply. The degree of precision and accuracy required for each sample should also be considered, along with sample stability relating to the time of travel from water supply to analysing laboratory, which for some remote supplies could be significant. It may be that, for some parameters at least, field based test kits could be used to provide adequate verification of the safety of a drinking water supply. Such test kits can yield almost instantaneous results, and although the level of accuracy may not match that of traditional laboratory analysis in all cases, this may be outweighed by increased number of tests it is possible to perform for the same cost and the benefit of being able to respond immediately to correct any quality issues that are indicated.

The frequency of sampling can be difficult to assess and may depend on resources and any existing regulatory framework. The WHO publication *Guidelines for Drinking-water Quality, Surveillance and control of Community Supplies – 2nd Edition*, gives well considered thought to sampling frequency and parameters which should be considered.

The results of analysis should be recorded, ideally so that they can be assessed over time to see improvements or deterioration of water quality. Correlation with any online quality monitoring is useful here, although such technology is beyond the means of many small water supplies.

The WSP should identify different tasks and roles, for example:

- Who will interpret analytical results?
- Who will be advised of the results and how will the information be disseminated?
- What actions should be taken if samples fail standards or are of concern?
- What emergency procedures are in place to respond to sample failures, such as the provision of alternative supplies, particularly to vulnerable individuals?

2.4.4 Section 4 – Action plan

Carrying out a risk assessment on a drinking water supply can identify a wide range of risks! It is very important that there is a prioritisation of risks, so that the wood can be seen from the trees. This can be done simply by transferring risks to an action plan – this can be done by writing lists of risks with scores and actions, or can be done electronically on spreadsheets (Table 2.3). However it is done, it is important that the WSP team agree on the prioritisation, the actions to be taken, who will carry out these actions, and the date that the actions will be completed.

Table 2.3 Example action plan.

Number	Risk Score	Action	Action Owner	Date for Completion	Comments	Complete (Y/N)
Action 1	128	Farmhouse lead pipework from mains to kitchen tap to be replaced June 2015. Farmer unable to afford to replace lead to bathrooms, will be done September 2016.	Farmer	June 2015	Advice given by health officer to householders to use bottled water, particularly important for children. Householders aware when pipework to kitchen replaced, teeth should be cleaned in kitchen.	N
Action 2	32	Discuss the possibility upgrade of cast iron network pipes at next water supply on 12/5/15.	L. MacLeod (chair of group)	12/5/15	No known bursts, but group aware of age and rusting. Cost likely to be main issue.	

It should be stressed that not all improvements will need capital investment. It is also recognised that bringing small supplies up to standard may take time, and resources may not immediately be available to carry out remedial works identified. There may be many which are operational or maintenance tasks, and there may be some interim tasks or procedures which are identified to deal with anticipated issues. for example, the WSP team may identify that there are significant problems with cast iron mains, that they are prone to bursts, and that there are discolouration events after bursts which make the supply unusable for consumers. While the best option may be to replace the mains, there may be no funding to allow for this. So, as an interim alternative, plans could be made for known risks, such as what should be done in the event of a burst on the system. For example, a survey could be done to determine whether appropriate flushing points are available, contractors' details should be available for mains repairs, etc, to allow a rapid response to any failures of the mains.

2.4.5 Section 5 – Review of the WSP

It should be stressed that WSP should evolve over time; they should be reviewed and updated regularly to keep them current and valid. They may also begin very simply and then become more comprehensive in time, as knowledge and resources become available and the process becomes embedded in the way the supply is managed. WSP should be viewed as ongoing management tools. It may be that not all information is available to allow a complete assessment of a supply, for example the condition of pipes in the network may not been known. In this case, the WSP should be completed as far as possible, and should be added to, expanded as further information is made available.

It is recognised that many small water supply owners, users and operators simply do not have the funding and resources that utility companies have, but the emphasis where this is the case is that as much should be done as possible with the resources that are available. While capital expenditure may be often be needed to fully mitigate risk, often operational interventions can minimise risk until funding can be raised. The WSP can help to identify these and be a useful reference to store all relevant information or references to the information, such as procedures for maintaining treatment, or instructions and likely frequencies for clearing leaves and plant debris from intakes.

2.5 PRACTICAL GUIDANCE FOR RISK ASSESSMENTS

2.5.1 The catchment and the source

A water supply ***catchment*** *is the area of land where the water is gathered, for example the fields on which the rain falls, and the burns and streams that flow from the area. The* ***source*** *is the water that has been gathered from the catchment, and is used to feed into a drinking water supply.*

Understanding and managing the risks from catchments and sources is critical to protecting the quality and quantity of the supply (Figure 2.2). The list below gives guidance on what should be assessed for the supply catchment and source. **The list below is not exhaustive** – further risks can be added.

- Identify any sources or potential sources of pollution. Meet with the land owners and users if possible, or take advice from the appropriate environmental or water supply enforcing agency.
- Where possible, take advice from any organisations which hold information on the geology of the area and any environmental water sampling. For example, environmental regulators may sample groundwater and may have data on nitrate and naturally occurring metal levels in the supply.
- Consider any issues with availability of water and any history of drought, year round and seasonal, and where necessary ensure that abstraction licenses are available and appropriate. Document the maximum possible output of water from the source where possible – this will be needed if further connections to the water system are needed.
- Identify any maintenance tasks, for example clearing leaves from surface water intakes.
- Identify any critical equipment and determine whether spares are needed. For example, if the supply relies on a borehole pump, the following should be checked:
    - Is the pump maintained in accordance with manufacturer's instructions?
    - Is there a spare pump in case of pump failure?
    - Where would spares for the pump be ordered from and how long would it take for them to be delivered?
    - Which engineers would be able to rapidly make repairs to the pump?

Figure 2.2 Assessment of risk in the catchment.

2.5.2 Water treatment

The need for treatment will depend on the quality of the raw water, and perhaps on the end user of the water. Treatment systems for small supplies are often different to those used by water utilities. Whereas utility companies have skilled staff and a wide range of expertise, often small supply treatment systems can be very basic, if they need to be easy to operate and maintain, require minimal skill, be cost effective and must have minimal health and safety issues. Figure 2.3 shows risk assessment of a treatment process.

Figure 2.3 Risk assessment of treatment processes.

Some points to consider are:

- It is important that all the treatment processes are clearly labelled; treatment vessels can all look very similar, particularly to the inexperienced eye. As well as causing water treatment problems, if treatment stages are confused with one another there may be health and safety problems, for example if the wrong chemicals are added to treatment stages, and the direction of flow through the process should be clearly marked on pipework and treatment vessels. The reason for the need for treatment should be understood and documented.
- There should be clear operating and maintenance instructions for all parts of the treatment system – if these are not available the manufacturer of the treatment should be contacted for advice.
- Operational and maintenance schedules should be prepared and maintained – and a record of the work completed should be kept. It is very useful to have a diary system to set out what tasks need to be tackled and when, and for completed tasks to be recorded. Understand what operational tasks are needed and what the frequency of the tasks should be in order to keep the treatment working efficiently and keep the drinking water supply safe. Operational tasks are actions that need to be taken to ensure that treatment processes operate in the way that they are designed. For example, this could be adding sodium hypochlorite solution into treatment tanks so that the system has disinfectant.
- The contact details of treatment manufacturers, contractors and suppliers should be maintained;
- If seasonal treatment is available, for example for manganese removal from surface water, everyone responsible for operating the system must be aware of the triggers for switching on the treatment;
- Advice should be taken from water treatment professionals if the treatment processes are not understood; this may need to be on an ongoing basis;
- Emergency procedures, contacts and information should be considered in advance of an emergency situation, and should be updated following a review of responses to any incidents. Emergency information could include, for example, details of resources and professional knowledge, from where contractors for pipe bursts can be contacted, treatment consumables such as UV bulbs can be ordered, to local government or regulator contacts for support. It can also identify the need for emergency procedures, for example how to deal with loss of supply.
- The capacity of the system should be assessed and documented to ensure that treatment systems are not overloaded, and so that any future expansion of the water supply system takes account of maximum flows through the treatment system.

2.5.3 Networks and storage tanks

Pipework and storage can contaminate drinking water supplies in a number of ways. Fractured pipes or leaking storage tanks, or animal access (e.g. rodents) can allow microbiological contamination. Pipe or tank materials can affect water quality, for example corrosion of iron pipes can generate particulate iron in the supply and discolour the water, and lead from pipes and tanks can dissolve into drinking water.

- If there are any plans showing the locations of pipework and tanks, copy them and store them with the WSP. If new pipes or tanks are installed on the system, ensure that the plans are updated so that there is an accurate record. Include details of pipe materials if possible.
- Talk to land owners and farmers to gain access to tanks or any other assets such as valves on the supply system, and to gain local knowledge. Give contact details of those responsible for the supply so that landowners and farmers can get in touch if they notice issues with the supply, such as bursts pipes or damage to tanks.
- If there is concern for the structural condition of assets, take professional advice.

- Inspect storage tanks, such as that shown in Figure 2.4. Be aware of risks from infiltration of surface water through roofs and sides of tanks; where the condition of tanks is poor or is unknown, consider engaging specialist contractors to inspect the tank and check its integrity. Inspect tank cleaning records if they are available and devise an appropriate cleaning schedule.
- Document the history of the network and any known weaknesses in the system and ways of managing the system through experience. Often dealing with emergency situations give a great deal about networks and how they can be dealt with. For example, maps should be marked showing where there have been bursts on the system, and pipe materials and conditions where they are known should be documented.
- The position of valves, hydrants and flushing points should be marked on maps or diagrams, and where possible and appropriate they should be inspected and tested to ensure they are able to operate in advance of an emergency situation.
- Where available, operating instructions for the network should be documented, using the experience of anyone who knows and understands the system.

Figure 2.4 Small supply collection tank.

2.5.4 Domestic distribution systems

The domestic distribution system (DDS) refers to the pipework, fittings and appliances between the distribution network and the taps used for drinking water for human consumption. The DDS is generally the responsibility of the owner of the property. In a very small water supply this may be one and the same as the network described in 2.4.3. The main issues in DDS are materials of construction and inappropriate storage conditions; legionella control is not considered in this document. Areas to consider are:

- The knowledge of householders, property owners and any maintenance staff should be considered, and where possible, plans of the building showing the positions of pipework and tanks should be viewed.

- Lead pipes, joints and tanks, copper from pipes and fittings, and nickel and zinc from fittings. This is particularly problematic if the supply is allowed to warm up in the DDS, if the water is in contact with the metals for long periods of time, or if the water has a low pH and is soft.
- Storage within buildings can cause microbiological deterioration of supplies, particularly if there is a low turnover of water, perhaps caused by oversized tanks or low residency in buildings, or if it is stored next to hot water supplies and is allowed to warm up.
- The security of tanks within buildings should be assessed. It is not unheard of for birds and animals to fall into drinking water tanks and decompose, so it is important that tanks are securely covered and made vermin proof.
- Cleaning schedules for tanks should be considered.
- Sampling of drinking water supplies should generally be at the point of consumption, such as the kitchen tap, and the condition, hygienic condition and suitability for disinfecting sampling points should be considered.
- If there is point of use treatment, then this is likely to be the responsibility of the householder. Those responsible for this household treatment should be advised to ensure that the treatment is appropriately installed and maintained;
- Consideration should be given to good plumbing practice, for example the risk of back-siphonage should be assessed, particularly where the quality of supplies can deteriorate within a DDS. Where there are activities which present a high risk to contaminating water, such as animal watering troughs, appropriate protection such as air gaps or non-return valves are necessary to protect the supply.
- Where there is any recycled water or rainwater systems which are separate to the drinking water system, it is very important that the systems are kept separate and have appropriate back flow devices to prevent contamination of the domestic system or the entire supply.

KEY POINTS

1. A drinking water safety plan (WSP) is a useful tool for assessing and managing risks on water supplies.
2. Development of a WSP does not need to be complex or costly.
3. Extensive guidance on the preparation of a WSP is provided by the World Health Organisation.
4. The best results come from involving a team of people with an interest in the supply.
5. The WSP should consider all risks from source through to where the water is consumed.
6. An action plan should be created to document how risks will be managed.
7. WSPs must be living documents and should be reviewed regularly.

REFERENCES

Scottish Executive (2006). Private Water Supplies Technical Manual. Scottish Executive, Edinburgh, UK. Available at http://www.privatewatersupplies.gov.uk/private_water/files/Full%20Doc.pdf (accessed June 2015).

World Health Organisation (1997). Guidelines for Drinking-water Quality 2nd edition, volume 3, Surveillance and Control of Community Supplies. WHO, Geneva, Switzerland.

World Health Organisation (2004). Guidelines for Drinking-water Quality 3rd edition, volume 1. WHO, Geneva, Switzerland.

Chapter 3

Establishing the metals problem: Risk assessment, sampling and analysis

3.1 INTRODUCTION

Metal contamination of drinking water may be determined by appearance or taste, but generally not by smell. Some metals may pose risks to health whereas others can cause nuisance, such as the staining of sanitary-ware or laundry, and a few are considered to be beneficial constituents of drinking water. The concentrations of potential concern are generally in the part-per-million (mg/l) to part-per-billion (μg/l) range and can be difficult to quantify without expensive analytical equipment. Sampling procedures can be critical in determining metal contamination and must be selected with care to avoid misleading results.

3.2 THE RANGE OF POSSIBLE PROBLEMS

Metals and metalloids are commonly encountered in the aqueous environment, with wide ranging impacts that include:

Impact on human health

The most conspicuous concerns relate to lead from old lead pipes and arsenic from contaminated ground water, particularly those ground waters in volcanic regions. Acute lead poisoning from drinking water is unlikely, most concerns relating to chronic exposure and neuro-toxicological effects, of which reduction in the IQ of young infants has received most attention (WHO, 2010, 2011). The toxicity of arsenic to human health ranges from skin lesions to various forms of cancer and toxicity depends on which arsenic species is present (Ferrante *et al.*, 2014). Other metals of potential health concern are copper (gastro-intestinal distress), chromium VI (stomach cancer) and manganese (neurological effects); although cadmium and mercury are highly toxic, they are not normally encountered in drinking water at significant concentrations. Many metals are needed for human metabolism in trace amounts, whereas magnesium (and possibly calcium) has protective effects in relation to cardio-vascular disease (Ferrante *et al.*, 2014).

Aesthetic effects and nuisance

The commonest problems arise from iron, either from the corrosion of iron water mains and pipes, from naturally ferruginous ground waters or from naturally contaminated surface waters, particularly upland lakes in which iron can be chelated by the humic and fulvic acids that leach from bog-land (Postawa & Hayes, 2013). Iron contaminated drinking water can have a yellow, red or brown colouration and is generally unacceptable to consumers. In the UK, iron discoloured drinking water is regarded as unfit for human consumption and water utilities have been prosecuted for supplying such water. Manganese is also a common natural contaminant of anoxic ground waters, is found in upland lake waters due to chelation, and can occur in deep reservoir and lake waters following thermal stratification. It has a tendency to plate out throughout a water system and cause black staining. Other aesthetic/nuisance impacts can be (i) green or blue staining of sanitary-ware and laundry by copper, arising from copper pipes; and (ii) a 'sandy' appearance due to particulate zinc, arising from galvanised iron pipes or the dezincification of brass fittings.

To establish if there is a problem requires a holistic approach, in line with drinking water safety planning approaches, but the reality of small and particularly very small water supplies is the frequent lack of available technical expertise and resource constraints. In Europe there are about two million small water supplies (Hulsmann, 2005) with a similar number in the United States. Small water supplies are also common in other parts of the world.

It is unrealistic to expect that the operators of small water supplies will be capable of much more than the simplest diagnostic assessments of metal problems. The principal burden for dealing with small water supplies must therefore rest with the local authorities or agencies that have responsibility for public health protection and housing standards. In practice, consumers, operators and agencies must work together, suggesting a framework of risk assessment procedures linked to educational programmes. The aim of this Chapter is to provide a basis for identifying problems with metals in small drinking water supply systems.

3.3 METAL SOLUBILITY AND INFLUENCING FACTORS

Oxidative states of various metals

Many metals exist in different oxidative states, each with their own solubility characteristics. The principal oxidants that determine metal solubility are oxygen and chlorine. The commonest circumstances that are encountered are:

- Iron in natural waters that are devoid of oxygen will likely be in the divalent ferrous state, which has a high solubility. When such waters become oxygenated, or come into contact with chlorine (frequently used to disinfect drinking water) the ferrous iron oxidises to the much less soluble trivalent ferric state which precipitates as a solid. Loose iron precipitates cause the water to become discoloured.
- Manganese in natural waters that are devoid of oxygen will likely be in the divalent manganous state, which has a high solubility. When such waters become oxygenated, or come into contact with chlorine (frequently used to disinfect drinking water) the manganous iron oxidises to the much less soluble tetravalent manganic state which precipitates as a solid that often sticks to the inner surfaces of pipes and sanitary-ware.
- In most cases, the corrosion deposits from lead pipes stabilise as divalent lead carbonate which is sparingly soluble, but sufficiently so to be of potential concern. If the lead pipes are in contact with persistently high free chlorine (particularly >2 mg/l) the lead carbonate can oxidise to the tetravalent lead dioxide, which is much less soluble (Hayes, 2010; Hayes, 2012).

pH, alkalinity and temperature

Metals have a higher solubility at lower pH values, particularly when the pH is less than 6, which is slightly acidic. At pH values above 9, it is possible to dissolve zinc out of brass, in some circumstances. As a general rule, in Europe, the preferred pH range is 7.5 to 8.5 for minimizing metal dissolution; in the United

States, higher pH conditions are often used. The impact of pH is much greater when the water has a low alkalinity (a measure of bicarbonate and carbonate) less than 50 mg/l (as $CaCO_3$). As a general rule, metal dissolution increases when the water has a higher temperature, typically by a factor of two comparing summer to winter conditions; the exception is copper which tends to passivate better in hot water systems.

Organic complexion

In upland areas, water that stagnates in peaty bog-land picks up significant concentrations of humic and fulvic acids, which are the degradation products of organic matter. These acids are polymeric, long-chain negatively charged molecules that can chelate (bind with and solubilise) a range of metals. The water that drains from such areas often has a straw colouration, has a pH less than 6 and an alkalinity less than 10 mg/l (as $CaCO_3$); such waters are highly corrosive to metals. Organic enrichment also occurs in lowland areas due to the discharge of treated sewage effluent and algal growth in standing water bodies. There is clear evidence that both types of organic enrichment can increase corrosivity (Hayes, 2010; Cardew, 2009).

Interaction between corrosion scales and water

The corrosion deposits that build up in water mains and premise plumbing are complex and the release of metals to water is influenced by many factors. Generally, the greater the contact time between the water mains or plumbing with the water, the greater is the tendency for metals to dissolve. The higher the flow rate of the water the more likely are physical scouring effects, also dependent on the physical condition of the deposits (ie: loose or hard). Corrosion scales can also be affected by the constituents of the water, such as treatment chemical residuals, manganese deposition and organics. In some cases, microbial activity may also be significant, particularly in the corrosion of iron water mains.

Corrosion inhibitors

In large municipal water supply systems, it is common to dose corrosion inhibitors. Orthophosphate can significantly reduce lead solubility, whereas silicates and polyphosphates can chelate ferrous iron, thereby reducing discolouration problems. The dosing of small water systems is much less practicable.

3.4 RISK ASSESSMENT OF SMALL WATER SUPPLY SYSTEMS

Introduction

In the UK, risk assessment of small, privately owned water supplies is the responsibility of the local authority. All supplies serving more than one home must be assessed. The relevant UK regulations (UK Secretary of State, 2010) do not extend to supplies to a single home. Elsewhere in Europe, most countries have none or very little information on the number of small supplies and even less on the very small supplies within their territory (Hulsmann, 2005); water supplies serving less than 50 persons or providing less than 10 m^3/day can be exempted from the provisions of the Drinking Water Directive. At least one in ten Europeans (40 to 50 million people) receive their daily drinking water from small and very small supplies, including private wells. In the US, non-community systems are not covered by the provisions of the Safe Drinking Water Act. It can be readily concluded that the extent to which regulatory provisions apply varies considerably.

Many of the smaller water supplies derive from streams, springs, shallow wells and even roof-water, all of which are susceptible to contamination. In Europe (Hulsmann, 2005), the commonest problems were reported to be microbiological contamination, nitrates, fluoride, chloride, ammonia, pesticides, radon, the metals aluminium, iron and manganese, and arsenic; no information on copper and lead was reported, most likely because of sampling limitations.

Approaches to risk assessment

Best practice must begin with the relevant Governmental departments for Health and for the Environment. It must also recognise the significant constraints, both knowledge and resources, that impede public health protection for the consumers of small water supply systems within their territory.

A tiered approach is could be:

- *Larger Supplies to >5000 People*: These could be regulated in the same way as municipal systems and subject to a Drinking Water Safety Plan, in the manner described by the WHO (2011), and be fully accountable to the relevant *National or Regional* agencies. Annual water quality reports and risk assessments could be published. Educational material could be available and promoted through health-care organizations.
- *Medium Sized Supplies to >500 but <5000 People*: These could be regulated in the same way as systems supplying >5000 people, the same drinking water quality standards could apply and they could be fully accountable to the relevant *National or Regional* agencies. Risk assessments may necessarily be truncated to focus on the more common threats to water quality. Water quality appraisal and reporting could be reduced to once every two years. Educational material should be available and promoted through health-care organizations.
- *Small Supplies to >50 but <500 People*: The same drinking water quality standards should apply and they should be fully accountable to the relevant *local* agency, under the guidance of the National or Regional agency. Risk assessments may necessarily be truncated to focus on the more common threats to water quality. Water quality appraisal and reporting could be reduced to once every five years. Educational material could be available and promoted through health-care organizations.
- *Very Small Supplies to <50 People*: The same drinking water quality standards could apply as guidelines. Within a District, a Cluster of very small water suppliers should be formed to strengthen shared knowledge and to enable collaborative awareness training. The Cluster could account to the relevant *local* agency, under the guidance of the National or Regional agency. Risk assessments may necessarily be much more abbreviated to focus on the most obvious threats to water quality. Water quality appraisal and reporting could be reduced to once every ten years. Educational material could be available to the Cluster from health-care organizations and promoted directly to all homes.
- *Institutional Buildings*: For all sizes of water supply systems, a risk assessment could be undertaken every five years at every school, day-care centre, health care centre and hospital. This requirement should be a duty of care placed upon building managers, who could be accountable to the relevant *local* agency, under the guidance of the National or Regional agency.

In practice, the risk assessment and reporting regime will necessarily have to reflect local circumstances. As an example, privately owned water supplies in England are assessed and reported by local authorities to the Drinking Water Inspectorate as follows:

- Supplies to >5,000 people or >1,000 m^3 per day are handled in the same way as public water supplies and reported annually.
- Supplies to 50 to 5,000 people or 10 to 1,000 m^3 per day are reported annually, regardless of the use to which the water is put.

- Supplies to <50 people or <10 m^3 per day are reported annually where water is used for a commercial or public activity, including food production.
- Supplies to <50 people or <10 m^3 per day are reported annually, but with less information, where water is used only for domestic purposes in two or more premises.
- Supplies that serve a single dwelling, and where water is used only for domestic purposes, are exempt.

Source water pollution by metals

Every source of drinking water has a catchment, also known as a gathering ground. Sources from large catchments may pose more risks, particularly in lowland areas with one or more population centres and industry. Large catchments in mountainous areas may pose fewer risks. An inventory of the catchment should include: (i) current or historic mine workings, (ii) quarries, (iii) industrial sites, (iv) forestry, (v) military bases, (vi) discharges from sewage treatment plants, (vii) proximity to roads, railways and airports, (viii) agricultural activities, (ix) population centres and (x) the nature of the water source(s) being exploited and their hydro-geology. Risk assessment for metals will likely be undertaken within a broader and more holistic assessment framework. In volcanic regions, ground water sources should be checked for arsenic contamination.

General water quality considerations

As discussed in section 3.3, the pH, alkalinity and organic quality require particular attention. A simple risk assessment matrix for possible problems with metals is shown in Table 3.1 below.

Table 3.1 Risk assessment matrix.

Parameter	High Risk	Moderate Risk	Low Risk
pH	<6	6 to 7.5	7.5 to 9
Alkalinity (mg/l $CaCO_3$)	<10	10 to 50	>50
Total organic carbon (mg/l)	>5	2 to 5	<2
Organic colour (° Hazen)	>10	2 to 10	<2

Inventory of metal components and potential sources

A wide range of metallic materials are used in a water supply system, including pumps, valves, tanks, water mains, meters, stop-cocks, internal plumbing and faucets. Common sources of metal contamination are:

- Lead from lead service lines and lead pipes in homes;
- Lead from brass components and brass pipes;
- Lead from solder with a high lead content (although this is mainly historic);
- Iron from cast-iron water mains;
- Iron and zinc from galvanised iron pipes and tanks;
- Copper from copper pipes;
- Nickel from chrome-nickel plated components, particularly faucets.

Additionally, metals may arise naturally (aluminium, iron, manganese) or as a consequence of using treatment chemicals (aluminium, iron).

An inventory will play a key role in identifying the need for mitigation measures and it will be helpful to take note of pipe lengths and diameters, as well as water flow characteristics to estimate water stagnation times. Intermittent use of pipes in institutional buildings may exacerbate metal release problems.

Visual evidence

The most conspicuous evidence will be staining of sanitary-ware, and possibly laundry. Green or blue staining is indicative of copper. Black staining is indicative of manganese. Yellow, red of brown discolouration is indicative of iron. Unfortunately, lead contamination cannot normally be discerned visually. Particulate zinc and lead may build up in the aerators attached to some faucets with a 'sandy' appearance.

3.5 SAMPLING AND ANALYSIS

Introduction

For more detail, the reader can refer to the Best Practice Guide on Sampling and Monitoring of Metals in Drinking Water (Postawa, 2012). In most circumstances, sampling and analysis should be undertaken by trained personnel.

Spot sampling

The simplest type of sample is a 'grab' or 'spot' sample; a sample bottle is simply filled with water at the point of sampling, typically a one litre volume. There are many limitations as the sample only represents the quality of the water at the instance at which it was taken. Variation in water quality at the point of sampling can only be determined by taking many more samples over a period of time; further, water quality varies from location to location. Sampling for metals at faucets is particularly challenging as variation arises from a range of factors, including pipe-work characteristics, flow rate, contact time between plumbing and water, temperature and fluctuations in water quality (particularly pH). It is not possible to characterise metal problems at an individual home by 'spot' sampling. However, random 'spot' sampling across a water supply system can determine the extent of metal problems if sufficient samples are taken (at least 100 and preferably >200) but there are obviously logistical limitations in small systems.

Split-flow composite sampling

Split-flow composite sampling can determine exposure to metals at an individual home and is illustrated in Figure 3.1. When water is drawn from the faucet for drinking or cooking purposes, a lever is switched in order to collect a small side-stream that can be collected over a day or week. The bulked sample can determine the average concentration of the metals ingested. This sampling technique is logistically onerous and can only be justified in the forensic examination of a home, in response to a health concern.

Sequential profile sampling

Lead release from an individual plumbing system can be investigated by:

- Flushing the pipe-work for several minutes;
- *Allowing the water to stand for a period of time*: 30 minutes should be adequate for gauging plumbosolvency condition, whereas over-night standing can indicate potential peak concentrations;
- Taking 12 × 1 litre samples in sequence.

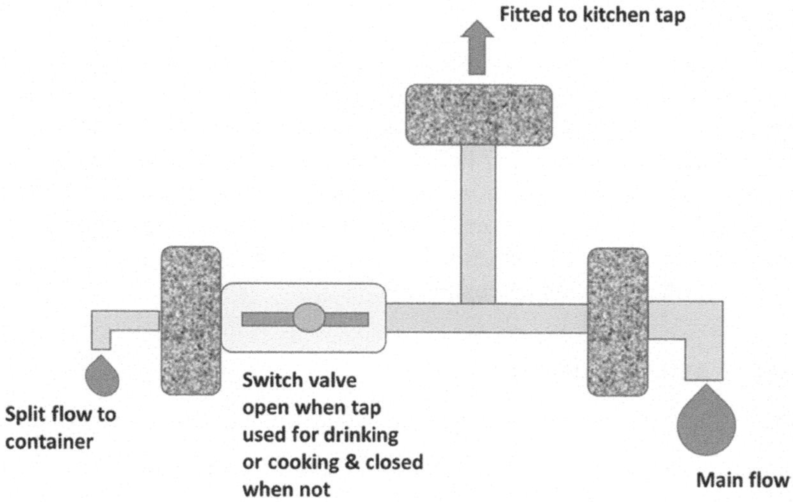

Figure 3.1 Diagram of a split-flow sampling device (Postawa, 2012).

The results can be used for diagnostic purposes and are illustrated in Figure 3.2 for samples obtained from Edmonton in Alberta, Canada (Hayes & Croft, 2014) for two stagnation periods: 30 minutes and over-night standing (6+ hours). The peak concentrations relate to water stood in a lead service line and provided an indication of the plumbosolvency of the water supplies in the system, albeit variations from home to home were found to be significant. This technique has less value for investigating other metals.

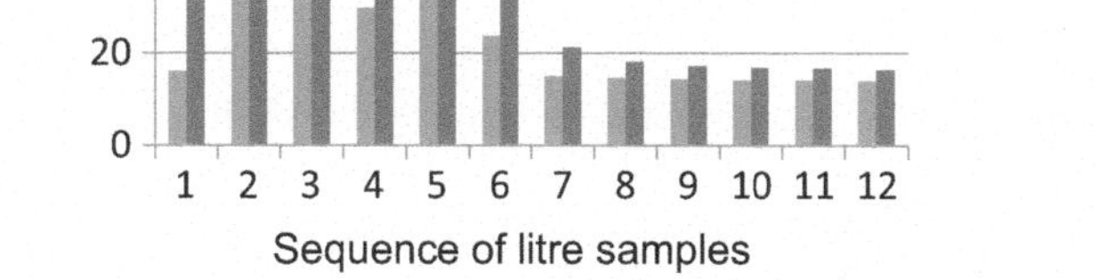

Figure 3.2 Results of sequential sampling.

Differentiation between particulate and dissolved metals

It is possible to differentiate between particulate and dissolved metals by filtration through a 0.45 µ membrane, as commonly used in bacterial analysis.

Analysing for metals and pH

It is preferable to use an accredited laboratory that uses recognised and approved methods. The National or Regional agency should be able to advise. The most commonly used method involves Inductively Coupled Plasma – Mass Spectroscopy (ICP-MS) which has very low detection limits (Postawa, 2012). The costs involved limit the amount of work that can be undertaken in a small water supply system. However, field testing methods for several metals have recently been developed, using catalytic DNA sensors that fluoresce in the presence of a target substance (www.hach.com). Tests are performed by taking a 1 mL water sample and can detect cadmium, copper, lead, mercury, uranium and zinc. To measure pH, simple colorimetric paper strips can provide an approximation.

3.6 CONSUMER AWARENESS

Educational campaigns have an integral role to play in public health protection and should encompass:

- Simple fact sheets that explain drinking water from 'source to tap' and how to undertake basic risk assessments;
- Fact sheets on the mitigation measures that can be implemented;
- Specific information sheets on arsenic and lead to promote awareness;
- Contact lists that provide details for all the relevant agencies;
- Access to a Community Chat-Room or Blog to encourage knowledge dissemination.

KEY POINTS

1. Some metals are of health concern whilst others cause nuisance.
2. The oxidative state of the metal determines its solubility.
3. Solubility is also strongly influenced by pH, alkalinity and organic quality.
4. Risk assessment should review catchment characteristics, general water quality and the occurrence of metallic components.
5. Sampling methods have limitations and must be selected with care.
6. Field testing methods are now available.
7. Consumer awareness has an integral part to play in public health protection.

REFERENCES

Cardew P. T. (2009). The impact of ortho-phosphate treatment on lead in drinking water in North West of England. Proceedings of International Conference, COST Action 637, October 2008, Lisbon.

Ferrante M., Oliveri Conti G., Rasic-Milutinovic Z. and Jovanovic D. (2014). Health Effects of Metals and Related Substances in Drinking Water. IWA Publishing, London.

Hayes C. and Croft T. (2014). Optimisation of plumbosolvency control using computational modelling techniques: a demonstration project for the Alberta Government, working with the City of Calgary and EPCOR (Edmonton). Project Report to the Alberta Government. Contract Number: 140100. February 2014.

Hayes C. (2010). Best Practice Guide on the Control of Lead in Drinking Water. IWA Publishing.

Hayes C. (2012). Code of Practice on the Internal Corrosion Control of Water Supply Systems. Specialist Group on Metals and Related Substances in Drinking Water. IWA Publishing, London.

Hulsmann A. (2005). Small systems large problems. A European inventory of small water systems and associated problems. WEKNOW/ENDWARE.

Postawa A. (2012). Best Practice Guide on Sampling and Monitoring of Metals in Drinking Water. IWA Publishing, London.

Postawa A. & Hayes C. (2013). Best Practice Guide on the Control of Iron and Manganese in Water Supply. IWA Publishing, London.

UK Secretary of State (2010). The Water Supply Regulations 2010 (England and Wales). The Stationery Office Ltd., London.

World Health Organization (2010). Booklet on Childhood Lead Poisoning. WHO Document Production Services, Geneva, Switzerland. Available at http://www.who.int/ceh/publications/leadguidance.pdf (accessed June 2015).

World Health Organization (2011). Guidelines for Drinking Water Quality, 4th edition. WHO, Geneva, Switzerland.

Chapter 4

Sources of metals in small water supplies

4.1 ORIGIN OF CONTAMINANTS

It is likely that for the majority of small supplies where metals are an issue, the contamination is present at source. This likelihood will depend on the particular metal that has been detected – arsenic and manganese are likely to be derived from raw waters, while lead, copper and nickel are more likely to be (but not necessarily) derived from metals used in plumbing materials. Iron can originate both at the source or from ferrous pipework and fittings used in the supply.

In order to deal effectively with source contamination of a small supply it is necessary to have a clear understanding of the location and characteristics of the source – a requirement applying to all types of contamination, not just metals. This understanding is not always present – the exact location of an abstraction point from a surface water source may not be known and documented, or the characteristics of an underground source and the principal horizons from which water is drawn may not be clear. The latter situation is especially common, and it may be necessary to seek professional advice to determine borehole characteristics. Such advice comes at a cost and this may be prohibitive to some owners and users of small water supplies. Where tracer testing is undertaken, extreme care should be taken not to contaminate the supply or others nearby. Even where a source is thought to be underground, it may still be influenced by surface water containing significant concentrations of metals.

Where multiple sources are present, each may present a similar or very different risk of contamination. The risk from each source should be assessed separately, and each sampled individually, ideally on multiple occasions and under a range of conditions. In some cases it may be that a significant proportion of the risk from contamination can be addressed by de-commissioning one of the multiple sources and only using those that present a much lesser risk, where water resources allow. This will often be the simplest and cheapest solution to the problem, however the ability to choose between sources of differing quality is something of a luxury, especially in locations where water is scarce. The benefit of a risk-based approach such as a water safety plan especially lends itself to this type of supply where the risks and mitigation measures for each source need careful and systematic assessment and documentation.

Contamination and the risk of contamination can change significantly over time, either increasing or decreasing, necessitating a thorough knowledge of catchment activities and characteristics which is regularly reviewed and updated. Once again the use of a regularly updated water safety plan to manage and document risk and mitigation measures will prove invaluable. Activities in the catchment can introduce a new source of contamination or exacerbate an existing one. Common examples encountered are harvesting of forestry trees or changes in agricultural practices which lead to leaching from soils and mobilisation of metal contaminants. Manganese and aluminium are frequently liberated in this way and can cause problems in previously unaffected supplies.

4.2 CONTAMINATION OF SURFACE WATERS

Surface waters will pick up metals from the environment that they have passed through upstream. Metals will readily leach from soils and rocks and enter water supplies, especially where conditions exist to favour the existence of metallic elements in soluble form. Acidic conditions, perhaps due to naturally occurring organic acids or sulphide deposits, will greatly increase the leaching of metal ions into water courses. Complex interactions between other mineral ions will cause precipitation and re-dissolution of metals. Insoluble metals may be transported in sediments, either as minerals or in native form. Surface waters may also have an input from groundwater as springs and other subterranean sources augment the total flow, adding metals derived from the local geology and soil deposits.

Modes of transport are complex and beyond the scope of this publication, however it is helpful to consider these in general terms when attempting to risk assess the supply. A geological map covering the whole catchment of the watercourse may be useful in indicating the kinds of metals that may be present in a supply. Even more useful may be analysis of stream sediments and water trace chemistry, datasets of which are produced and published in some countries by organisations such as the national environmental regulator or geological organisation.

Certain metallic minerals are commonly associated with specific rocks or type of rock. These are summarised in the Table 4.1.

Table 4.1 Geological sources of common metals.

Metal	Typical Geological Association	Example Rock
Arsenic	Mineralisation (often sulphide) around igneous rocks. Also associated with iron and manganese.	Granite
Cadmium	Finer sedimentary rocks; sulphide minerals.	Mudstones
Chromium	Basic or ultrabasic rocks, clays and fine sediments	Olivine, Peridotite, Gabbro, some Basalt
Lead	Acid igneous rocks; some sedimentary rocks containing organic material. Also secondary mineralisation.	Granite, some clays and shales
Nickel	Basic or ultrabasic rocks, clays and fine sediments. Some sedimentary rocks derived from basic igneous sources. Also sulphide mineralisation, often associated with iron and manganese.	Some greywackes
Uranium	Igneous rocks; organic rich sedimentary rocks; hydrothermal mineralisation	Granite, black shales

Concentrations of some metals in surface waters can vary temporally as well as spatially. Manganese is a good example of this, and many water supplies where manganese is present have a manganese 'season' when concentrations are at their peak. This can be due to a number of different environmental factors and mechanisms, with each supply having slightly different characteristics. One notable factor governing manganese concentrations in lake-fed surface waters is seasonal ' turnover' which can occur towards the end of the year as temperatures drop. This can create an imbalance in lake stratification, whereby denser, cool water begins to form on the surface overlying warmer, less dense water at depth. As this deeper water has gradually become more depleted in oxygen over the Summer months, any manganese present in sediments will have converted to its reduced, soluble form and dissolved. At some point the imbalance in water densities becomes too large and the water body 'turns over' bringing high manganese concentrations to the surface where they can drawn off into water supplies.

Other seasonal peaks maybe due to seasonal rainfall changes mobilising soluble manganese from soils surrounding water bodies. Consequently, to fully understand the likely metal content in a water supply and likely treatment requirements it is worth undertaking sampling at a range of times through the year and under different climatic conditions. Often this is not done with small water supplies for various reasons, including cost, meaning that decisions on treatment and risk mitigation are often taken on the basis of very limited data.

Not all contamination is naturally occurring of course. In some areas, man-made contamination can be the main source of metals in drinking waters, sometimes at concentrations which give cause for concern. Often these are point sources, meaning that water concentrations can vary greatly from one location to the next. Heavy industrial sites (current and disused), landfill sites, and all activities associated with mining and mineral processing can be significant sources of metal in water sources. Some common industrial sources of metals in surface waters are listed in Table 4.2, but this is by no means exhaustive.

Table 4.2 Common man-made sources of metals in water supplies.

Metal	Common Man-Made Sources
Arsenic	Smelting, wood preserving, manufacture and use of pesticides, sheep dipping
Cadmium	Metal plating, pigments, batteries
Chromium	Metal plating, leather tanning, chemical production, pigments, petrol refining
Lead	Fuel production, lead smelting and mining
Manganese	Batteries, fertilisers, metal foundries, leather and textiles
Mercury	Chemicals, paints, electrical equipment, dentistry, fungicides
Nickel	Steel production, electroplating, ceramics, batteries and glass, petrol refining
Tin	PCB manufacturing, marine paints
Vanadium	Metal manufacture, petrol refining
Zinc	Galvanising

Local topographical maps can yield useful information when risk assessing water sources as they can indicate potential point sources of metal contamination. Current and disused sources of metals may be shown on the map.

In particular, any mining activity, past or present, should be viewed with suspicion when identifying potential sources of metals in water supplies. This is firstly because for economically viable mining to have taken place there must be significant deposits of naturally occurring minerals and potentially higher concentrations in local water bodies, and secondly because the mining process itself and associated mineral processing can release significant concentrations of metals into the aquatic environment, in some cases for many years after mining activity has ceased. Spoil heaps and tailings ponds can contain extremely high concentrations of metals, which can contaminate watercourses both gradually and in single, high impact events for example during periods of flooding.

Water emerging from abandoned mines can be a rich source of metal contamination due to the condition known as acid mine drainage where water acidified by sulphide minerals such as iron pyrite leaches metals from the rocks, often with the aid of microbial action. Mines affected by this need not be limited to those mining metalliferous ores. Coal mines often contain iron sulphide minerals which oxidise and leach iron and other metals in the acidic water draining from the mine (Figure 4.1).

Figure 4.1 Iron hydroxide precipitate in a small watercourse.

Some sources of metal contamination can be very localised, making them hard to detect or predict. This can be particularly true of waste materials which have been disposed of inappropriately. Examples include unexpectedly high lead concentrations in one source despite the apparent absence of any lead piping. It was subsequently discovered that a number of large lead-acid batteries had been buried near that source many years ago and were continuing to leach lead into the supply. Where the source of increased metal concentrations is not clear, systematic sampling of the upstream water course can be one of the best ways of locating the origin. Multiple water samples run for a range of metals using inductively coupled plasma analysis techniques (ICP) should not prove too expensive, especially when compared against the cost of prolonged investigations or the installation of unnecessary treatment processes.

4.3 CONTAMINATION OF GROUND WATER

Groundwater, where available, is often selected for small supplies because it tends to be reliable, both in terms of quantity and quality. Although initial construction costs may be higher than for a simple surface water abstraction due to the expense of drilling, on-going treatment costs may be significantly less, making this an attractive option. It should not, however automatically be assumed that all groundwater derived sources will be free from all contamination, including metals. Apparent groundwater sources can be influenced by surface waters, including any contaminants they contain. Limestone karst areas are especially vulnerable to surface contamination, sometimes from a considerable distance away as sinkhole and cave networks can convey water very efficiently across large distances. Other geological scenarios where water flow is through fractured rock rather than confined aquifers can suffer from surface water influence. Some small supplies that are assumed to be groundwater sources, turn out, on closer investigation, to be wholly surface derived with all the contamination risks this implies. Boreholes, springs and wells that are not properly capped and maintained also risk contamination from direct infiltration of water at the point at which the groundwater comes to the surface (Figures 4.2 and 4.3).

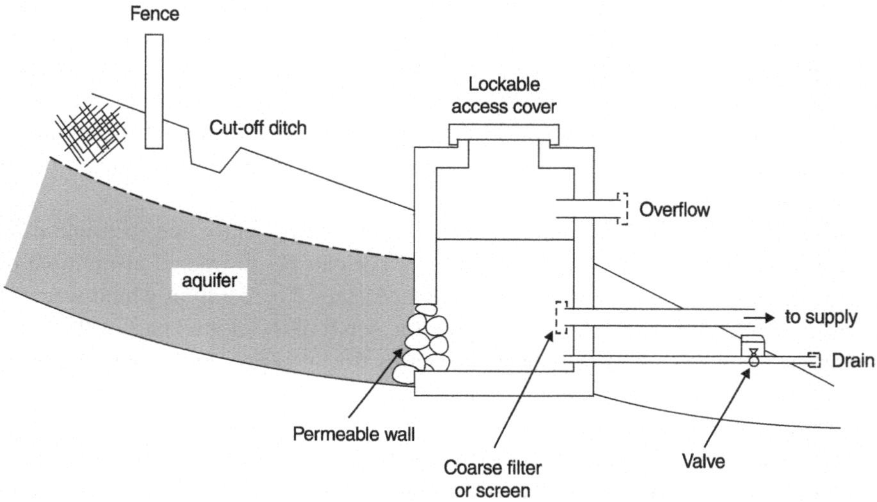

Figure 4.2 Diagram showing well constructed spring source, designed to prevent the ingress of contamination (Private water supply technical manual, DWQR, 2006).

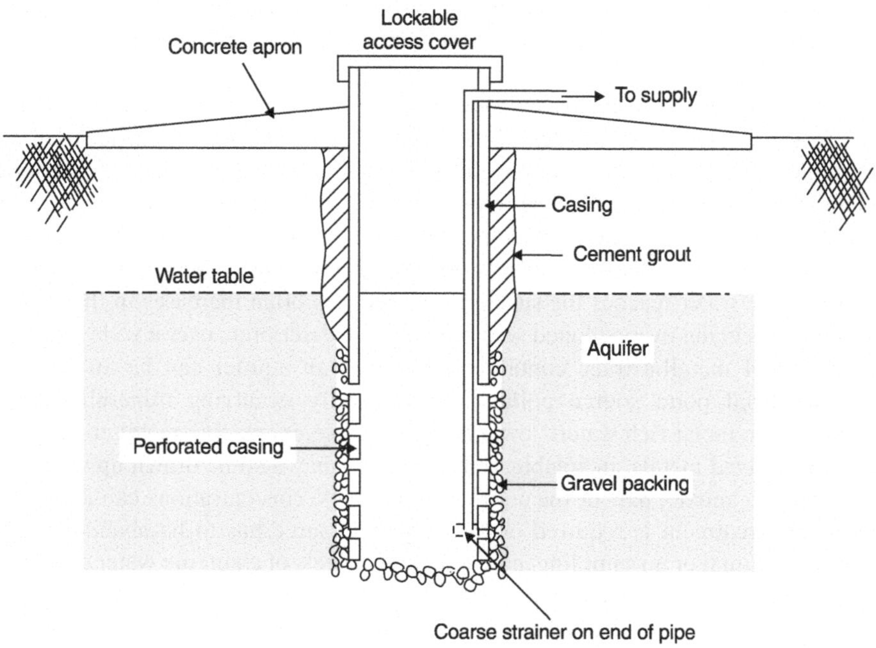

Figure 4.3 Section through well constructed and sealed borehole headworks (Private water supply technical manual, DWQR, 2006).

Aquifers themselves can become contaminated by man-made pollution and this can include metals. As with surface water sources, groundwater abstraction from locations close to current or disused industrial sites such as factories, mines and waste disposal facilities should be viewed with suspicion and investigated further. It would be prudent to commission expert advice and undertake metals sampling if there is a possibility that metals could have leached into groundwater from sources such as these. Where contamination is present it is likely that it will be preferable to locate an alternative source rather than to attempt to treat or remove the contaminant. It should be noted that owners of boreholes and other groundwater sources have a moral and, often, legal responsibility to ensure that their borehole does not introduce contaminants into an aquifer. Responsible construction and decommissioning procedures should be followed in order to ensure this.

Of course, groundwater will often contain naturally occurring metals due to mineralisation in the rocks from which it is abstracted. In some boreholes, water can be abstracted at multiple depths, with each geological horizon having different mineral characteristics. The depth at which water in boreholes originates can be controlled to some extent by varying the depth of the slotted part of the borehole casing which allows water to enter from a specific horizon. In this way, it is potentially possible for boreholes in close geographical proximity to be abstracting from different geological horizons and therefore yield water with very different quality characteristics.

When planning a new groundwater source, expert geological advice should be sought, both in terms of the likely quantity of water yielded, but also the quality. Other similar groundwater sources nearby may also provide an indication of potential quality, but this is not necessarily accurate, for the reason described above and because mineralogy can vary spatially. This can be especially true where the trace metals are derived from superficial deposits, such as those deposited by glacial or fluvial action. In many cases it will only be possible to determine the exact chemical properties of water from a groundwater source once it has been developed and analytical samples taken. Having said this, where all similar supplies in an area require treatment to remove a certain metal it is highly likely that any new source will also require it.

Concentrations of metals in groundwater sources may vary with time. This may follow seasonal patterns, due to seasonal recharge of aquifers, or it may be due to longer term changes. These can occur both naturally or due to the effects of the pumping itself. An example of this is where groundwater pumping begins to deplete oxygen rich waters near the top of the aquifer (Figure 4.4). Over time, less oxygenated water from deeper in the aquifer can be drawn into the borehole. Being lower in dissolved oxygen it can contain higher concentrations of dissolved metals, for example iron and manganese, which may begin to precipitate out when the water reaches the surface. Salinity also often increases in these boreholes, and where a whole aquifer is being overexploited water quality can deteriorate over a wide area.

Lateral movement of metallic trace contaminants within an aquifer can be an issue. These may originate from industrial point source pollution or naturally occurring mineralisation. Prolonged pumping can draw these metal rich waters towards the borehole, increasing metal concentrations in the water abstracted. Dissolved metals, in soluble, reduced, form may also be drawn upwards by continued pumping from a deeper, anoxic, part of the aquifer. Sometimes, concentrations can increase to such an extent that additional treatment is required or the borehole source has to be abandoned. Many small water supplies have minimal or no sampling, and unless the effects of changing water quality are visually apparent, they may not always be identified in time to prevent adverse health effects. For this reason, groundwater quality monitoring is a valuable tool which should be included in any management plan for a small water supply.

There are a number of naturally occurring bacteria which thrive in anaerobic conditions and derive energy for survival by oxidising soluble iron (II) to insoluble iron (III). A similar group of bacteria also exist which thrive on manganese. These organisms can be found in springs and streams where anaerobic conditions exist and are often characterised by slimy orange or brown deposits (black in the case of manganese) and sometimes a swampy or musty taste and odour to the water. These bacteria can colonise boreholes, which can lead to problems such as clogging of the borehole, decreased pumping efficiency and even pump burnout. Once established, colonising iron bacteria are difficult to remove – infected wells can be cleaned physically followed by chemical treatment or pasteurisation. Prevention is better than cure, and scrupulous hygiene procedures in developing and maintaining the borehole should help to prevent it becoming contaminated.

Figure 4.4 Wellhead in Malawi.

The choice of materials used in the construction of a borehole is important in protecting water quality, as these in themselves can leach metals into the water, especially where natural characteristics of the water favour this. Borehole casings, rising mains and pumps should all be constructed from an appropriate type and grade of material for use in the specific circumstances (Figure 4.5). Even small quantities of an inappropriate metallic material can cause concentrations in drinking water to exceed regulatory or health-based limits. Some countries maintain lists of approved products for use in drinking water applications, and these should be used wherever possible. A number of European countries produce lists of products and materials that have been tested and approved for use with drinking water; in the UK, the Drinking Water Inspectorate (DWI) and the Water Regulations Advisory Service (WRAS) maintain lists and in the United States, the National Science Foundation (NSF) maintains a comprehensive list of certified products that are considered safe to use in contact with drinking water. Often, but not always, these approvals lists mainly contain products more suited to larger, municipal water supplies and it can be difficult to identify approved products for use in small supplies.

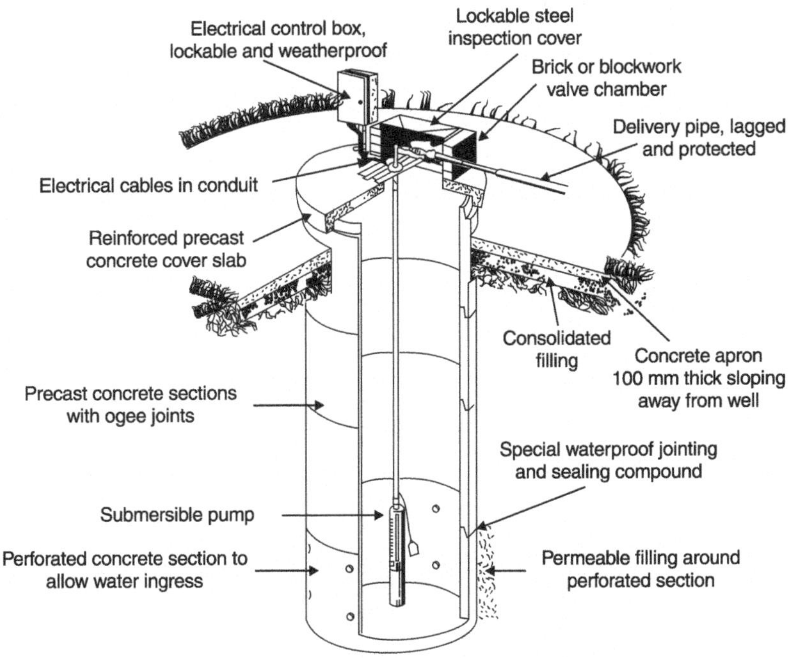

Figure 4.5 Diagram of typical borehole construction (Private water supply technical manual, DWQR, 2006).

Regardless of material used, all equipment has a finite life and may leach metals into the water as they corrode (Figure 4.6). With groundwater systems, equipment can often be out of sight, making it hard to detect early signs of corrosion. This makes it especially important to implement regular maintenance schedules on such equipment.

Figure 4.6 Badly corroded borehole pump.

4.4 CONTAMINATION FROM TREATMENT PROCESSES

Treatment processes themselves can be sources of metals in small water supplies, both by design and unintentionally. As with equipment at water sources, there is a need to ensure that materials and processes appropriate for use with drinking water are used, and the information in Section 4.2 on approved products lists applies equally here.

This section will briefly consider the ways by which treatment processes commonly used on small supplies can introduce metals into the water.

4.4.1 Ion exchange processes

Ion exchange is a process by which contaminating substances in the water are substituted for ions of another, less undesirable, substance. Nitrate and water hardness the most common substances removed in this way, but a number of ion exchange resins are available and may be used to remove a range of substances, both cationic and anionic with a reasonable degree of selectivity. The process is based on the principal that the undesirable ions are replace by another species which is less problematic. Often this species is sodium, which is very soluble, and therefore largely invisible to the consumer. At low concentrations it is reasonably benign, but where a large amount of sodium is added by the process, in excess of approximately 200 mg/l this may give rise to a taste that is unacceptable to the consumer. This aesthetic threshold forms the basis of the regulatory standard in some parts of the world, including Europe.

4.4.2 Filter media

Filter media of various types are often used in the treatment of small water supplies, depending on requirements (Figure 4.7). These can range from simple sand media to activated carbons and metallic oxides. Once again, the importance of using media approved for use with drinking water is strongly advised, with particular attention paid to any manufacturer recommendations for commissioning and flushing prior to use. This is vital, as even the most simple media can leach substances, including metals that had been adsorbed onto the media. Aluminium is especially prone to leaching from inadequately washed media. Where a supply suddenly encounters issues with elevated metal concentrations and new filters or filter media have recently been installed, inadequate flushing on commissioning is often the cause.

Sloughing or re-dissolution of accumulated metal deposits can also be in issue where filters are used on a supply. This can occur for a number of reasons:

- Change in quality of incoming water, especially pH
- Lack of maintenance resulting in a build-up of filtered deposits and eventual sloughing
- pH correction filters causing elevation of pH and precipitation of metals onto media with eventual sloughing

Many of these issues can be overcome by adequate pre-treatment, correct design and sizing and appropriate backwashing facilities (combined with use at the correct frequency and flow). If the supply is intermittent, as is not uncommon with small supplies, and the filters are sitting wet for long periods, chemistry across the filter can change, often with a reduction in pH, causing dissolved metals to be flushed into the supply upon start up. A run to waste facility can be useful in such cases.

Figure 4.7 A loose media filter on a small water supply, used for iron removal.

4.4.3 Oxidation

Where a treatment process changes the oxidation state of naturally occurring metals in solution in the water, this can result in them becoming insoluble and precipitating out of solution. Of course, this can be an intentional part of the treatment process, for example where iron and manganese are brought out of solution using an oxidising agent such as chlorine or oxygen prior to removal via filtration. This process needs to be tightly controlled to ensure that metals are precipitated in the correct part of the system to ensure their removal.

Unintended oxidation can occur, perhaps where turbulent flow aerates the water or where chlorine is added for disinfection purposes and this can cause metals to precipitate and accumulate in locations in the treatment process or downstream where they can be re-suspended and cause problems. A common example of this on small water supplies is the situation where pH correction filters are used at the end of the treatment process to raise the pH of soft waters. As the pH is elevated in the filter media, even small concentrations of metals present in the water can precipitate onto the filter media and become concentrated there. Changes in flow or water chemistry can suddenly release these in to the supply at concentrations hugely in excess of those occurring naturally in the water, causing problems with aesthetics, regulatory compliance and, potentially, the health of people consuming the water. The issue can be controlled through good system design and, especially, by ensuring pH correction filters have the ability to be backwashed if metals precipitation is likely to occur.

4.4.4 Coagulated treatment processes

Coagulation, flocculation and clarification, whereby positively charged metal ions such as aluminium and iron are used to neutralise negatively charged particulate matter in the water and cause it to settle out are commonly used processes to treat larger municipal water supplies. The need for chemical dosing, careful process monitoring and control (preferably online), and a large treatment footprint makes this a process that is far less common on small water supplies, and consequently probably beyond the scope of this publication. Nonetheless, it is possible that it may be used in places and that it could be a significant contributor to the metal content of the water where the process is not properly managed. Aluminium in particular has a narrow pH window at which it is insoluble in water and able to be removed by downstream filtration. If water pH is allowed to deviate outside this window, soluble aluminium can pass through the treatment process and achieve high concentrations relatively quickly.

4.4.5 Corrosion of metals within the treatment process

Corroding metal fittings and pipework within the treatment process can contribute metals into the water. Iron is probably the most common example of this, but lead, chromium and nickel could also occur where they are present in materials in contact with the water being treatment. Any corrosion may be especially acute where soft water is present prior to pH correction or where corrosive treatment chemicals are dosed into the water. The issue of corrosion as a source of metals is considered in the next section and later in this book.

4.4.6 Metal contamination from treatment chemicals

Small water supplies tend to avoid the added complexity of chemical treatment due to lack of expertise and ability to adequately monitor and manage the associated risks. Having said this it is worth noting that many chemicals used in the water treatment process are acidic and corrosive, and therefore may dissolve metals where they are stored incorrectly, going on to contaminate water supplies where they are later used.

4.5 CONTAMINATION IN DISTRIBUTION PIPEWORK

It is well known that pipeline materials can contribute metallic substances to drinking water. Considerable investment is undertaken on large municipal water supplies in order to counter the effect of corroding iron mains and accumulated deposits of other metals such as manganese, through renewal of water mains and removal of deposits via flushing or some other technique.

Small water supplies are equally prone to such contamination. Although distribution system lengths may be shorter, a number of factors often occur to make the issue equally, if not more, acute:

- *Inadequate water treatment to remove metals:* This may lead to accumulations of metals and related compounds at points of low velocity in the distribution system that can later be re-suspended following a flow disturbance to greatly impair the quality of water supplied to the user. Manganese is probably the most common metal to accumulate in this way – it often takes the form of a very fine black oxide which coats inner surfaces of pipework and can cause very dark brown or black water and stained fittings. Inadequately removed or overdosed metal-based flocculant chemicals, although less commonly used in the treatment of very small water supplies, can also cause problems later in the system. The two most commonly used are based on iron, which takes the form of brown or red sediment and aluminium which often manifests itself as a colourless or white precipitate, often gelatinous in nature.
- *Use of inappropriate distribution materials:* It has already been identified that in many locations, regulation of small water supplies is weak or absent altogether. In most parts of the world, larger

water supplies have tight controls placed upon them regarding the fitting and use of products which have been thoroughly tested to ensure they do not leach substances into the water they are carrying. Small water systems often do not benefit from such regulation, and furthermore, are often designed and constructed with minimal expenditure and low understanding of the technical issues and risks. Systems may be wholly or partly constructed from materials that are completely inappropriate for that use, and which may, either continuously or under certain conditions, be altering the chemical characteristics of the water passing through them. The effect of such contamination may be obvious, in terms of a visible taint, taste or odour to the water, or it may be completely insidious and continue indefinitely, unsuspected by the individuals consuming the water. Both scenarios can have significant implications for the health of individuals concerned.

- *Inadequate conditioning of water:* The conditioning of water, that is the balancing of pH and dissolved mineral content to reduce the corrosivity of the water and its tendency to dissolve materials and fittings in the distribution system, is an important but often neglected part of water treatment. Never is this more so than on small water systems where such processes are often completely absent and awareness is poor. The topic is a highly complex one, which is dealt with more fully in Chapter 7 of this guide, but the issue is especially acute with naturally soft waters of low pH and dissolved mineral content. Aside from issues with the premature corrosion and failure of pipes and fittings, cases of water supplied with concentrations of plumbing metals far in excess of the regulatory and health-based standards have been found. This issue is greatly exacerbated where, as discussed previously, non-approved plumbing materials are present which are not suitable for use with potable water. Implications for human health in such circumstances can be very serious indeed.
- *Poorly understood distribution system:* Often, routes of pipework and locations of intermediate storage tanks are not well understood (Figure 4.8). System plans are often absent and their production should be a priority when any Water Safety Plan is produced for a system. This lack of understanding can lead to potential sources of contamination such as areas of poor turnover, inappropriate materials and cross connections not being identified.

Figure 4.8 Raw water tank on small water supply.

4.6 CONTAMINATION FROM PLUMBING FITTINGS

Most of the issues identified for distributions systems apply equally to plumbing systems inside buildings. Some are even more acute – the complexity of pipework in older, larger premises can make investigations into the source of contamination even harder to spot.

The risk of inappropriate materials is even greater here – the range of plumbing materials and fittings available is extensive, and control over quality in most countries is comparatively weak. It is often not obvious where fittings contain metals likely to leach into water and even materials with some sort of accreditation or quality mark may be permitted to contain substances that have the potential to cause problems. Nickel, copper, zinc and lead are the most common materials that are leached out of pipes and fittings – the extent of potential aesthetic and health effects varies with the concentration, but quite high concentrations can be achieved via exposure to relatively small lengths of problem plumbing.

Leaching from domestic plumbing can be an issue for all water supplies, but is often more acute for smaller ones due to the issues already discussed – a lack of awareness and regulation in combination with inadequate water conditioning. Water characteristics such as pH, inorganic carbon, dissolved oxygen and mineral content are just some factors that influence metal solubility and small water supplies often exhibit limited control of these factors. These factors are discussed in more detail in Chapter 7.

KEY POINTS

1. Many metals in small water supplies are present in the source.
2. Metals can be naturally occurring or arise from man-made activities.
3. Careful choice of source location and construction of infrastructure can prevent or reduce contamination.
4. Treatment processes can introduce metals into water if inappropriate or poorly controlled.
5. Distribution systems and plumbing fittings can introduce metals post-treatment.
6. Water conditioning, good system design and careful choice of plumbing materials is key to reducing post-treatment contamination.

REFERENCES

Abesser C. and Robinson R. (2010). Mobilisation of iron and manganese from sediments of a scottish upland reservoir. *Journal of Limnology*, **69**(1), 42–53.

British Geological Survey. Regional Geochemical Atlas. www.BGS.ac.uk (accessed June 2015).

DWI. List of Approved Products for use in the Public Water Supply in the United Kingdom (2015). Drinking Water Inspectorate, London, UK. www.dwi.defra.gov.uk (accessed June 2015).

NSF. Certified Plumbing and Related Products. NSF, Ann Arbor, MI USA. www.NSF.org (accessed June 2015).

Scottish Executive (2006). Private Water Supplies Technical Manual. Scottish Executive, Edinburgh, UK. http://www.privatewatersupplies.gov.uk/private_water/files/Full%20Doc.pdf (accessed June 2015).

Sorg T., Schock M. and Lytle D. (1998). Leaching of Metals from Household Plumbing Materials – Impact of Home Water Softeners. National Risk Management Research Laboratory, USEPA, Cincinnati OH USA.

World Health Organisation (2004). Guidelines for Drinking Water Quality 3rd edition. World Health Organisation, Geneva, Switzerland.

WRAS Approvals Directory. Water Regulations Advisory Scheme, Crumlin, Gwent, UK. http://www.wras.co.uk/directory/default.htm (accessed June 2015).

Ziemacki G., Viviano G. and Merli F. (1989). Heavy metals: sources and environmental presence. *Annali dell'Istituto Superiore di Sanita*, **25**(3), 531–535.

Chapter 5

Water treatment processes available for use on small water supplies

5.1 PROCESS SELECTION

The range of treatment processes appropriate for use on small water supplies is often somewhat different to those available to the municipal water supplier. The ideal process for a small supply has the following characteristics:

- it is cheap to install, operate and maintain;
- it runs reliably with minimal monitoring;
- it requires no technical knowledge of the operator;
- it successfully treats water to the required standard under a variety of conditions and changing raw water qualities;
- minimal consumables are required, and no hazardous chemicals;
- no hazardous or difficult wastes are produced;
- the process has a small footprint and minimal power requirements;
- products are available 'off the shelf' and can be installed by the non-specialist.

Of course, no such process exists, nor is one ever likely to. The most successful treatment processes used on small supplies combine some of these attributes to some extent, but the selection of a treatment process is usually a compromise between competing pressures. Many of these pressures may not be readily apparent on initial investigation, and certainly not to the untrained eye. It is not uncommon for a treatment process installed on a small water supply to fail or operate sub-optimally due to a hitherto unappreciated factor or constraint. Consequently it is advisable to seek specialist advice when considering and installing treatment for a specific issue, even if it is intended that any process will subsequently operate with minimal intervention.

This chapter describes the various treatment options available, modes of operation and suitability for various applications. The table at the end of the chapter, in Section 5.2.6 serves as a summary and provides process engineering information including typical specifications and design constraints.

5.1.1 System installation and maintenance

Once the most appropriate treatment system has been selected care must be taken to ensure its installation has no detrimental effects on the water quality. This may seem rather obvious but it is not unusual to find equipment installed using copper or other metal pipework which can leach metals into the water. Various types of plastic pipework which are suitable for potable water use such as MDPE should be used when possible.

All equipment will require maintenance at some time and this should only be entrusted to personnel who have been appropriately trained to carry out the work in accordance with the manufactures instructions. It is not uncommon to find that inappropriate modifications are carried out to treatment plant which indicates a lack of understanding of the way equipment is designed to operate. The results are not hard to imagine.

5.1.2 Point of use versus point of entry

Unlike large municipal water supplies, small supplies often utilise treatment of the supply at or near the location where the water is to be used rather than treating the whole water supply close to its source and distributing treated water to users. This is for a number of reasons:

- there are multiple users of the source, but no coherent organisation to enable the water to be treated in one place and supplied to everyone;
- different users of the source may have differing requirements, financial constraints or views on the level of treatment required;
- the characteristics of any pipework and storage to transport water from the source may make it unsuitable for transporting treated water, rendering additional treatment necessary where the water is to be used ;
- Point of Entry (PoE)/Point of Use (PoU) treatment is marketed for smaller supplies and is often available 'off the shelf' and is simpler (or perceived to be simpler) to install, design and maintain than larger scale water treatment processes;
- users may feel that it is more efficient to treat the water to different standards depending on usage – for example only locations where water is going to be consumed may have disinfection.

PoE treatment treats the whole flow of water entering a property. Such treatment is usually located outside the property in an outbuilding, although it may be located inside the building at a point in the water system prior to consumption. It has the advantage of ensuring that all water entering the property is of an acceptable standard and water should be safe to drink no matter where in the house it is consumed. PoU treatment only deals with the flow to a particular outlet, which has advantages in terms of cost as the sizing of treatment equipment required to treat a proportion of the flow is lower. PoU may be acceptable where water is only consumed at one location in the property, but presents a clear risk if there is the potential for water from other, untreated, outlets in the property to be consumed, perhaps by those unaware of the risks.

Where metals are present in the source, the decision as to which type of treatment to install needs to consider a number of other factors beyond the immediate potability of the water. High concentrations of metals can cause unsightly staining of sanitary ware and even the hair of those washing in the untreated supply. Metals may precipitate out in pipework and tanks, and influence the corrosion of household plumbing.

The following factors may influence the choice between installing point of entry or point of use treatment on a water supply containing significant concentrations of metals:

- the concentration and nature of metals present;
- impact of the metals on sanitary fittings;
- where in the property water is likely to be consumed (Note that this should include activities such as tooth brushing and cooking);
- the number of water outlets in the property;

- space available for the installation of water treatment processes at PoE/PoU locations – note that this needs to include space for access and maintenance on any treatment installation;
- availability of power;
- hydraulic factors and requirements for tanks and pumping.

Where PoU treatment is installed on a water supply that does not meet the necessary quality standards, other taps in the property should be labelled as non-potable to prevent inadvertent consumption. Some countries discourage or forbid the use of PoU devices on waters that contain certain contaminants, and others set conditions of use. In some circumstances, for example arsenic removal, several PoU devices may be installed in series.

5.2 TYPES OF TREATMENT

The difference between PoU and PoE treatment is generally one of size – most of the treatment processes described in the section below may be found in either application.

5.2.1 Physical filtration

Loose media filters

These are filters consisting of a hollow metal, plastic or fibreglass shell filled with loose filter media, resting on a support medium, which is often gravel (Figure 5.1). They can be up-flow or down-flow and contain several different media layers. The particular media selection can be varied depending upon the particular contaminants requiring removal. Several filters may be installed in series in order to provide progressive removal of solid material and prevent clogging of subsequent filters or other treatment processes such as ultra-violet disinfection.

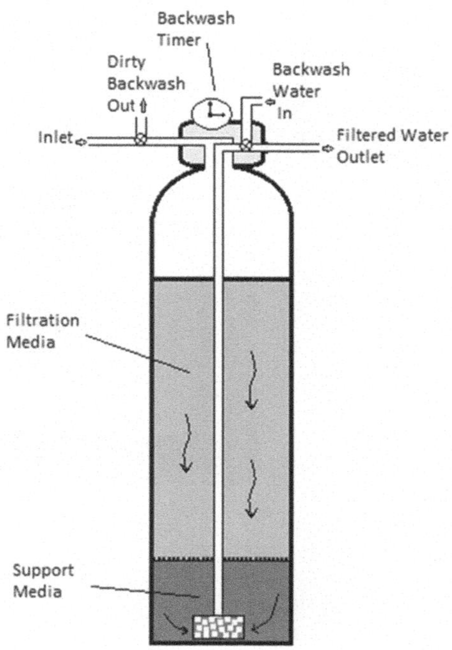

Figure 5.1 Loose media filter with backwash timer.

In order to achieve effective removal via filtration, it is necessary for the metal to be present in insoluble form as a suspension. In some cases this may be the case, and simple filtration will be sufficient, however it is common for a proportion, or most, of the metal to be present in reduced, soluble form. This will be especially true where the water originates from underground and is anoxic. Oxidation may be achieved in a number of ways. Iron may require little assistance and will oxidise readily on contact with the air. Manganese is more stable in reduced form and an oxidising chemical and/or catalyst such as chlorine or potassium permanganate may be required. Clearly, on small water supplies operated and maintained by non-specialists it will be desirable to use the simplest option and minimise hazards from chemical dosing. Some simple experimentation may be required to determine the optimal oxidation conditions. Clearly, it is desirable to maximise the amount of metal precipitating just prior to the filter – too soon and insoluble material could block pipework, too late and soluble metal could pass through the filter only to precipitate at a later stage and foul a more sensitive downstream treatment process with potentially serious consequences for any disinfection stage.

Although iron and manganese are the most common, other metals which may be removed by oxidation, precipitation and adsorption/filtration include chromium (III), selenium and arsenic (III). Often, a number of metals are present and co-precipitation can be highly effective. This is especially true for arsenic and iron, and where there is sufficient iron present arsenic will often be removed alongside.

Physical filters will need to be backwashed regularly in order to remove insoluble material that has accumulated on the filter. Inadequate backwashing will result in a rapid deterioration in filter performance and a corresponding loss in water quality and quantity. It is possible to install timer units (Figure 5.2) that will enable backwashes to be performed automatically, and these are to be recommended where the user of the supply has limited expertise or interest. Any backwash system requires an adequate source of clean water, and this needs to be considered when designing the system. Due to the size and complexity of these systems they are more suited to PoE supplies.

Figure 5.2 Typical loose media filter installed for iron removal, with close up of timer unit (right).

Cartridge filters

Pre-formed cartridge filters of all sizes are available to remove particulate material from water supplies. These may be constructed in a number of different ways and from various materials, including the following:

- Pleated paper or fibre;
- Plastics such as polypropylene or nylon;
- Cellulose;
- Membrane;
- Fibreglass;
- Ceramics.

Cartridge filters will be rated by the maximum size of particle they will allow through. A 'nominal' rating means that a stated percentage of pores in the filter are smaller than a specified size. An 'absolute' rating states the maximum pore size that may be found anywhere on the filter. Filters are often installed in series in descending order of pore size to maximise removal efficiency and protect downstream filters or other treatment processes.

Cartridge filters are convenient because they are easy, clean and quick to change even by non-experts (Figure 5.3). The range of sizes means that PoU installation is a viable option. Because they are not usually backwashed they are replaced once flow through the filter reduces to a certain value as pores become clogged with deposits. This aspect rather limits their suitability for metals removal, although they are sometimes used in circumstances where small concentrations of iron or manganese are present in oxidised (insoluble) form. In such cases a 5 micron filter would typically be used, although some trial and error may be needed. An upper limit of 300 microgrammes per litre of iron is often quoted as the upper cut-off above which cartridge filtration may not be effective and a fully back-washable media filter is required. The same is true where iron or manganese is in dissolved or colloidal form.

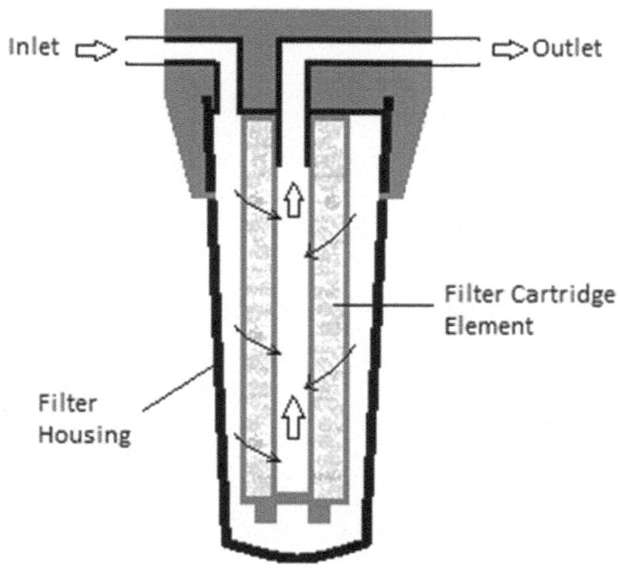

Figure 5.3 Simple diagram of cartridge filter.

Cartridge filters which contain some of the adsorptive media described below may also be available, and these may be useful for removing low concentrations of problem metals, even at point of use. Once again, the inability to backwash these filters and relatively low contact times limits the scope for these filters to be used in this application. Careful monitoring is required in order to identify when contaminant breakthrough is imminent and a filter requires changing. Figures 5.4, 5.5 and 5.6 show a range of cartridge filters and installations.

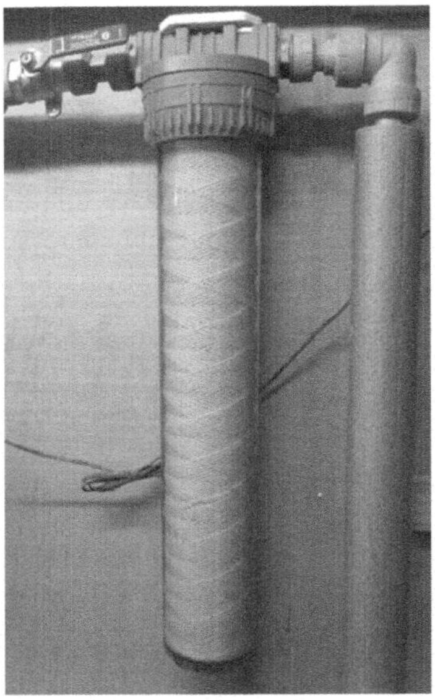

Figure 5.4 Wound fibre cartridge filter.

Figure 5.5 Point of entry cartridge filters installed to achieve progressive filtration at 100 and 5 micrometres.

Figure 5.6 Point of use treatment consisting of two cartridge filters upstream of an ultra-violet disinfection unit.

5.2.2 Chemical or adsorptive filters

As well as providing a physical filter, these filters can be adapted to contain media that chemically remove contaminants, either by catalysing a chemical reaction or enabling adsorption of the contaminant onto the surface of the media. Because the process involved is a chemical one, there may be less of a requirement to oxidise a metal contaminant to insoluble form. In fact, deposition of insoluble metal precipitates on these filters may impede performance and shorten the life of the filter. The main types of chemical filtration media are described below:

Granular Activated Carbon (GAC)

This substance is a common ingredient in many water filters and is derived from a range of natural materials, including wood, coconut shell, bone and coal. Different carbons have different removal characteristics and specialist advice should be sought. GAC is most commonly used to remove organic compounds but it can also be effective at removing certain metals from water, either by adsorbing them directly or catalysing their oxidation and enabling their subsequent removal. Removal of mercury, lead, iron, zinc, chromium, cadmium and copper have all been reported (Ahmedna *et al.*, 2004; Kadirvelu & Namasivayam, 2003), but metals removal across a filter may vary considerably between types of carbon and other factors such as water pH. Filter performance will deteriorate with time as the carbon becomes exhausted. This necessitates careful monitoring as increasing concentrations of metals in the filtered water may not always be obvious to the consumer. Where monitoring is not possible or unlikely to be carried out, these filters are not recommended. Activated carbon provides an ideal growing medium for microbes, and a poorly maintained GAC filter can present a health risk in its own right.

Manganese dioxide

The removal of manganese from water requires oxidation, and is greatly assisted by the presence of a catalyst media such as manganese dioxide. Both natural and manmade forms exist – the natural form is known as greensand. There is some evidence that manganese dioxide may also assist the oxidation and removal of arsenic (III) from water (Driehaus *et al.*, 1995).

Activated alumina

Activated alumina is a specially treated form of aluminium hydroxide which produces a durable, granular filter medium. Its best known use is for the removal of fluoride, at which it is extremely efficient, however it has also proven to be effective at adsorbing certain metals, most notably

- Arsenic (in arsenate form),
- Selenium,
- Beryllium,
- Chromium,
- Uranium.

The media is selective, and removal may be affected by the presence of other rival ions which are more preferentially adsorbed. The process is also highly pH sensitive, for example arsenic removal is optimal in a narrow window between pH 5.5 and 6 – clearly, chemical addition may be required to lower the pH to achieve this consistently, which puts this option beyond the scope of many domestic small water supplies. For many applications on small supplies, granular ferric hydroxide may be a more forgiving and less costly medium, even if the absorptive capacity of activated alumina can be greater under optimal conditions.

Activated alumina can be regenerated once the media has been exhausted – this is commonly done with sodium hydroxide, again making this a less suitable technology for the untrained operator. If not regenerated, spent media needs to be replaced, with the old media carefully disposed of as it may contain significant quantities of absorbed toxic metals. Because of its low cost, activated alumina is usually replaced and not regenerated in small systems. It is likely that pre-treatment will be required to remove sediment in the water which would otherwise foul the alumina filter.

Granular ferric oxide/hydroxide

Various iron-based media are available, and these appear to be showing promise as a single use media in removing a number of metal contaminants from water supplies. They are an established technology for arsenic removal and evidence suggests some effectiveness at removing the following metals:

- Antimony
- Cadmium
- Chromium
- Lead
- Molybdenum
- Nickel
- Selenium
- Vanadium

As with other adsorbent media, most waters are likely to require pre-treatment in the form of simple filtration in order to remove sediment and other particulate that would otherwise block the adsorption media. The performance of the media may vary according to the presence of other metals and anions such as phosphate, nitrate, chloride and silicate. As with all treatment technologies, specialist advice should be sought prior to undertaking significant expenditure.

Other media

A multitude of other filter media are available with specific performance characteristics. Some are a hybrid of those described above, while others are based on different substances including clays and minerals such as zeolite, activated glass and other metal oxides such as titanium and cerium. Once again, different media will be suited to different contaminants and circumstances and specialist advice should be sought.

Cartridge filters

Cartridge filters are available that have adsorptive properties in addition to providing physical filtration. Such filters may contain loose granular activated carbon within the cartridge or the filter material may be impregnated with carbon. The degree of chemical filtration achieved will depend on the nature and amount of absorptive material present as well as the water flow rate through the filter and the length of time it has been in use. It is likely that these filters will only be truly effective in circumstances where the extent of contamination is marginal. Careful monitoring will be required to identify when the adsorptive capacity of the filter has been exhausted.

5.2.3 Ion exchange

Ion exchange technologies can be useful for removing metals from small supplies in certain circumstances. Metal ions in solution in the water are substituted for another ion, often sodium, on the surface of the resin. The additional sodium is not normally a problem, except for individuals with certain health conditions. Ion exchange filters are unlikely to be a particularly cheap or simple solution but can be used to deal with some metals that are otherwise difficult to remove. Figure 5.7 shows a simplified, typical ion exchange process.

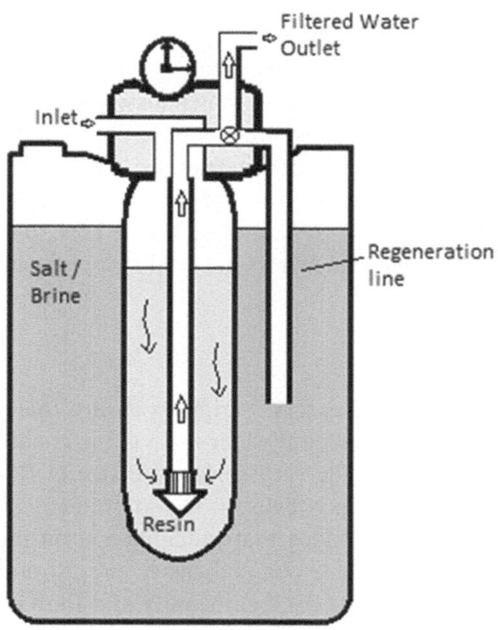

Figure 5.7 Small scale ion exchange unit, showing brine reservoir and regeneration timer.

For removal of most metals, cationic ion exchange resins are required. Some resins are highly selective and will mainly remove only specific metals. Others are more general, but will tend to exhibit a differential affinity, effectively creating a removal hierarchy. Where significant concentrations of a metal high on the affinity series are present, this will mean that this metal will be preferentially removed from the water over ions from metals lower down the series. This factor needs to be considered when designing an ion exchange process. Calcium and magnesium ions associated with water hardness are found quite high in the affinity series for many resins, meaning that water hardness can reduce the effectiveness of the ion exchange process for metals lower down the series. The presence of metals, such as iron and manganese, which oxidise to an insoluble form can cause fouling and increased head loss across the filter.

Before long, all the sacrificial sodium ions on the resin will have been replaced with cations from the incoming water, most of which will hopefully be the metal requiring removal. At this point, the resin will require regeneration by washing with a solution of sodium chloride. Some units are equipped with automatic regeneration equipment, and all the user needs to do is to ensure that there is sufficient salt in the unit. The unit containing the resin may be situated inside the larger brine tank, or it may be separate from it. Some smaller PoU ion exchange units are disposable and cannot be regenerated. In this case, the old, exhausted, resin cartridge is simply replaced with a fresh one. With all except fully automatic units, the user will need to establish when the media has become exhausted. Although some cartridge-based resins include a dye which changes colour on exhaustion, the need for regeneration won't always be obvious from the water emerging from the filter, so manufacturers'/installers' instructions should be followed very closely indeed.

For some metallic elements, anionic exchange resins can be effective, in particular for arsenic (V), uranium and selenium removal. These systems operate in a similar way to cationic systems, although they are regenerated using chloride ions.

5.2.4 Reverse osmosis membrane

A reverse osmosis (RO) water treatment system is one where water is forced through a semi-permeable membrane at high pressure, leaving contaminants behind in the concentrate water. The process can be high cost and have relatively high process water loss, but it may offer a solution to metals contamination where other treatment processes are not an option or have been unsuccessful. RO has been documented as removing, or reducing concentrations, of boron, lead, cadmium, chromium, copper, arsenic and antimony. The process has the additional advantage of also removing pathogens from the water – a factor which may justify any additional cost. It should be noted that nano-filtration membranes may also remove some metals in certain circumstances, with a reduced need for re-mineralisation.

Design of RO systems

Reverse osmosis treatment systems need careful sizing to ensure that sufficient water is available when it is required. This must take into account water lost via the concentrate stream, and, for point of entry systems, any available storage. The very small pore size of RO membranes mean that any solid material at all in the water will quickly cause it to foul, reducing throughput drastically. This could include any metal oxides or hydroxides that precipitate form the water. As a consequence, RO membranes should only be used where comprehensive pre-filtration has been installed. This may increase the size of the treatment process significantly and limit its use where there are space constraints, as is the case with many under–sink PoU applications (Figure 5.8). Some manufacturers supply small RO units that come with integral cartridge pre-filtration, although for waters with significant amounts of suspended solids, iron or manganese, it is possible that back-washable media filtration stage may be needed.

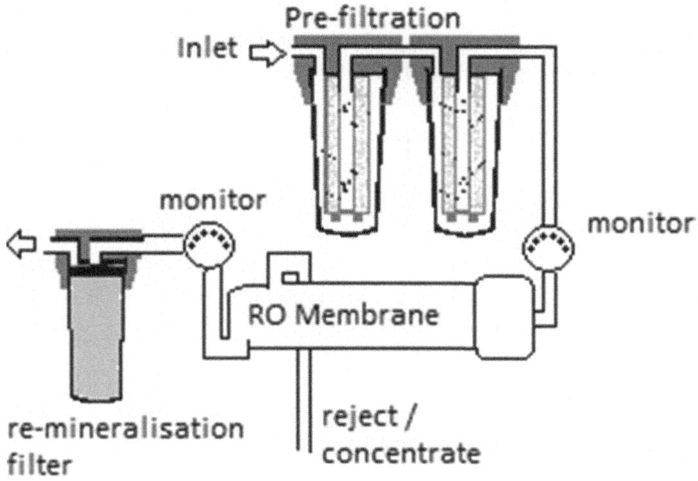

Figure 5.8 Diagram of typical small scale RO membrane installation showing pre-treatment and re-mineralisation filters.

As RO membranes effectively remove dissolved minerals from the water, the resulting permeate water will be low in dissolved solids and consequently highly aggressive to metal plumbing systems and fittings. The result can be leaks due to corrosion and staining of sanitary ware, and in extreme cases, hair washed in water containing dissolved copper. A simple post-membrane filter containing an alkaline medium such as crushed limestone should be sufficient to condition the water, although the optimal contact time within the filter will need to be calculated carefully.

Monitoring and maintenance

Part of any installation of an RO membrane should be a means of monitoring the integrity of the membrane. This could be done in a number of ways, but the monitoring of conductivity is probably the simplest. Any sudden increase in conductivity comparative to that in the incoming water could indicate failure of the membrane and the need to replace it. Although industrial RO membranes are routinely cleaned using chemicals, such maintenance is probably beyond the unskilled owner/operator. Unless a service contract with a competent specialist is available, it may be necessary to accept that the lifetime of a membrane without regular cleaning may be shorter.

5.2.5 Distillation

Distillation can be used to remove impurities, including metals, from water and several systems are available which can be used in point of use installations. They operate by heating water to boiling point so that pure water vapour carries over to the condenser, leaving any impurities behind in the boiling chamber. Systems are available in models which provide a continuous, if small, flow of water as well as batch systems where distilled water is stored until it is needed. The choice of system will largely depend on household consumption patterns and preference.

Distillation systems require a relatively high initial outlay and incur significant running costs due to the energy required to boil the water. They may provide a useful alternative for difficult to remove metals where other, less expensive options, are not appropriate. Of course, distillation systems also disinfect the water, meaning that it will be safe to drink, provided it is supplied and stored correctly. The systems are not appropriate where water contains volatile organic substances w hich will be carried over from the still with the water. Activated carbon filters may be used to remove these prior to the distiller. Some people find that distilled water tastes 'flat' or bland due to the low mineral content.

The waste liquid which remains in the boiler can be disposed of to sewer or by other means. In order to function efficiently, the boiler chamber will need to be periodically cleaned in accordance with manufacturer's recommendations. This is likely to require the use of chemicals, with cleaning frequencies depending on the mineral content of the water being treated.

5.2.6 Treatment options and specifications

Table 5.1 Water treatment process specifications.

Process	Typical Specifications	Typical hydraulic Loadings	Performance	Comments
Loose media filtration (gravity feed or under pressure)	Sand size 0.5 to 1.5 mm; Sand depth 0.5 to 1.0 m	Filtration rate 1 to 5 m/hr; Backwash rate up to 30 m/hr	Suitable for influent with turbidity <10 FTU. Reduction to <1 FTU possible. To achieve a lower turbidity of 0.1 FTU would be difficult without coagulation.	For higher influent turbidities need settlement before filtration. Settlement without coagulation may not give very good performance, as turbidity may be too fine to settle rapidly enough.
Cartridge filtration	Pleated paper, fibre, plastics, fibreglass, ceramics; Pore size up to 100 microns. 5 micron most commonly used, but can go down to 1 micron with implications for headloss.	Usually sized based on maximum recommended flowrate for specific cartridge models.	<1 FTU may be achievable depending on pore size.	Dispose of cartridge or clean by hand

(Continued)

Table 5.1 Water Treatment Process Specifications (*Continued*).

Process	Typical Specifications	Typical hydraulic Loadings	Performance	Comments
Granular activated carbon (GAC)	Carbon size 1 to 3 mm; Bed depth up to 1 m; Empty-bed contact time (EBCT) up to 30 mins; Surface area 1000 to 1500 m^2/g	Filtration rate 5 to 30 m/hr; Backwash rate up to 20 m/hr; Based on EBCT rather than filtration rate	Tastes and odours can be removed with bed-lives of up to 3 years before regeneration. Metals removal may not be significant long-term. Mn removal can occur.	Wide range of GAC types available, and suitability depends on specific applications and what needs to be removed. Cartridge filters are available.
Activated alumina (AA)	Particle size usually 1–3 mm; Surface area up to 300 m^2/g; Use BS EN 13753 specification for water treatment	Filtration rate 5 to 30 m/hr	Can give good performance depending on frequency of replacement or regeneration	Used to remove fluoride but also arsenic, selenium, beryllium, chromium and uranium. Iron coated AA available for some metals e.g. arsenic.
Iron based adsorption media such as granular ferric oxide	Surface area up to 350 m^2/g	Filtration rate 5 to 30 m/hr. But based on EBCT rather than filtration rate	Can give good performance depending on frequency of replacement, eg arsenic removal down to 4 μg/l	Used to remove arsenic and a wide range of other metals. Wide range of proprietary products available e.g. Bayoxide 33, Sorb 33, GFH
Ion exchange	Cation exchange or anion exchange resins with bed depth 0.2 to 1 m; Pressure vessels operate at about 0.6 MPa.	Typically around 2–3 minutes contact time; Flowrates up to 50 m/h	Good performance possible for many metal ions, depending on regeneration frequency. It is possible to reduce calcium and magnesium almost to zero. Need to regularly top up regeneration chemicals	Most commonly used to remove calcium and magnesium in water softeners. Routinely regenerated using sodium chloride. Some metals exist naturally as anionic complexes e.g. arsenic and uranium. Anion exchange resins capacity can be influenced by natural organic matter in the water, as this is removed.
Reverse osmosis	Pore size 0.1 to 1 nm; Pressure 10 to 70 bar	Usually expressed in terms of recoveries which may be from 10 to >80% depending on water quality, size and type of membrane.	Up to 99% removals are possible	Fouling may require pre-filtration. High energy consumption. Small domestic units produce only a few litres per day.

KEY POINTS

1. The choice of treatment process is usually a compromise between performance, practicalities and economics
2. Point of Entry/Point of Use Systems are often used to treat small water supplies
3. The appropriate treatment arrangement depends on individual circumstances and requirements
4. Loose media filters are often used to remove insoluble metals
5. Adsorptive filters can be used to remove certain metals in solution
6. Ion exchange resins may be available to selectively remove certain metals, but require regeneration
7. Reverse osmosis (RO) membranes will remove most metals but usually require pre-treatment

REFERENCES

Ahmedna M., Marshall W., Husseiny A., Rao R. and Goktepe I. (2004). The use of nutshell carbons in drinking water filters for removal of trace metals. *Water Research* **38**(2004), 1062–1068.

Driehaus W., Seith R. and Jekel M. (1995). Oxidation of arsenate(III) with manganese oxides in water treatment. *Water Research* **29**(1), January 1995, 297–305.

Ersoz M. and Barrott L. (2012). Best Practice Guide on Metals Removal from Drinking Water by Treatment. IWA Publishing.

Kadirvelu K. and Namasivayam C. (2003). Activated carbon from coconut coir pith as metal adsorbent: adsorption of Cd(II) from aqueous solution. *Advances in Environmental Research* **7**(2), January 2003, 471–478.

Chapter 6

Practical considerations of treatment for iron and manganese in small water supplies

Metals in water range from those that are generally considered to have purely aesthetic effects to those that are highly toxic. In practice common contaminants that are usually considered harmless can have serious negative effects on other treatment processes, particularly disinfection. This is particularly true with iron and manganese, two of the most common metals to affect small water supplies. It is vital to appreciate that there is no universal water treatment system that suits all scenarios and that a system that is appropriate for one particular site may be inappropriate for another in spite of apparently similar conditions. This chapter aims not only to deal comprehensively with the theory of treatment processes for these two metals, but also to provide an insight into the practicalities of operating treatment systems for these metals.

6.1 IRON

6.1.1 Principles of process design

By far the commonest metal considered as a nuisance is iron and although often regarded as one of the aesthetic parameters it cannot be ignored. How the iron appears in the water has been covered in Chapter 4. Apart from imparting (usually) a distinctive metallic taste to water it can deposit on sanitary ware, stain washing and cause severe problems with water treatment systems. These problems can occur with levels well below the EU indicator level of 200 micrograms per litre.

It must not be forgotten that water that appears aesthetically unattractive but is perfectly safe from a microbiological point of view may be shunned in preference for a source that appears clean but might be severely contaminated.

Iron occurs in several forms. These are ferrous, commonly called clearwater iron, ferric or rusty water and complexed iron. The nature of the complexes can be mineral or organic in nature. The latter are often referred to simply as organic and sometimes heme iron. Ferrous iron (Fe^{++} or Fe^{2+}) is found in boreholes, some wells and some lakes but all having anaerobic (anoxic) conditions. Ferrous iron changes to the insoluble ferric form (Fe^{+++} or Fe^{3+}) on contact with an oxidising agent, usually oxygen from the air. Mineral complexes can be found in any source but organic ones commonly appear in water originating in areas with high concentrations of peat. Although iron is usually a very visible and objectionable contaminant some complexes are very stable and they may only be revealed by analysis. One method used to reduce the adverse the effects of iron involves introducing a complexing or sequestering agent into the water. As this does not remove the iron this process is not considered further here.

It is important to ensure that any analysis of the water to be treated is relevant and representative of the water. This may seem to be rather obvious but it is quite common for samples to be taken from new sources such as boreholes without abstracting sufficient water to allow the conditions to stabilise. Samples may also be taken from a point of use without cognisance of any changes that may have occurred in the water between the sampling point and the source. As mentioned earlier ferrous iron can oxidise to the ferric form on exposure to the atmosphere and as a result it is not unusual to find part if not all the iron is in the ferric form by the time the analysis is conducted. This can result in analysis which indicates a low dissolved iron content but a high turbidity, neither of which are relevant to the point of treatment. It is vital to be aware of the state of the iron in the water when first drawn and therefore any analysis must check for total iron. It is preferable to carry out an onsite analysis for constituents which may change over time, including pH. Anyone who is involved with the supply and installation of iron removal plant should be equipped to carry out an appropriate on site analysis.

Apart from the amount of iron in the water and its form at point of treatment there are many parameters that can indicate whether or not the water is likely to be straightforward to treat. It is usually accepted that a minimum pH of 6.8 is required when the oxidant is oxygen but in practice some systems operate at an appreciably lower level. When an oxidant such as potassium permanganate is used a lower limit of 6.2 is often specified but this is possibly due to the potential for damage to the substrate of some filter medias. High levels of hydrogen sulphide, ammonia, sulphate, silica and phosphate may indicate a more difficult water, as might a total organic carbon in excess of 2 mg/l. Although the indications may be somewhat negative, collecting a sample from the point of treatment in a clear container, such as a polyethylene (PET) bottle, and watching what happens over a few hours can yield useful information to the experienced operator. Irrespective of all indicators the only way to guarantee that a particular treatment regime will provide a satisfactory solution is to conduct a pilot trial.

6.1.2 Iron removal – treatment methods

It is unusual to find iron concentrations reduced to undetectable levels by a small treatment plant hence complete removal may be an unrealistic expectation. It should, however, be feasible to achieve concentrations well within aesthetic and regulatory standards. If possible it is preferable to utilise a system that does not involve the purchase of chemicals.

Settlement

Probably the simplest way to reduce the level of iron is to detain the water in a cistern or series of cisterns where the ferrous iron will often oxidize to the insoluble ferric form. Frequently the ferric iron will precipitate allowing essentially iron free water to be decanted from the upper layer of water in the cistern. The cistern(s) can be elevated to give sufficient head for many purposes. This system is likely to be susceptible to unacceptable changes in temperature and microbiological contamination but maintenance is usually simple and limited to removing the accumulated precipitates. This can often be simplified further by installing one or more 'dump' valves.

It is essential to determine the detention time required for satisfactory operation bearing in mind the flow rate required and precipitation time. The 'PET' bottle test mentioned earlier is very useful.

Oxidation & filtration systems

It is often stated that the iron must firstly be converted to the ferric form to enable it to be mechanically filtered in a subsequent treatment stage. In order to convert ferrous iron to the ferric form, a minimum of 0.14 mg/l of oxygen is required per 1 mg/l of iron. To achieve this, water may be aerated by simply cascading into an essentially open cistern.

The time required for the iron to convert to a filterable form varies from water to water and may be measured in terms of minutes although more often in hours and sometimes weeks. Generally little attention seems to be given to this parameter with a result that, apart from after sufficiently long quiescent periods, the iron will enter the filter almost entirely in the ferrous form.

The simplest filters are similar in appearance to the settlement systems mentioned earlier but normally only one cistern/tank is employed. This is usually almost completely filled with filter sand with a minimum of around 10 centimetres of water above it. Water is usually distributed along the length of the filter bed using a length of pipe with holes or saw cuts in it. This is commonly referred to as a 'Sparge Pipe'. Filtered water is collected at the bottom of the vessel with a collection system commonly consisting of one or more pipes with slots cut in them. This is surrounded by one or more layers of material somewhat coarser than the sand to prevent the sand passing through the slots. The specific flow rate is usually quite small; around 0.5 m/hr is not unusual. It is important to note that in practice, even at these quite low flows the ferrous iron seldom has time to oxidise to the ferric form solely in the filter before it contacts the filter media.

When filter sand is employed as the filter medium it is usual for the system to have to 'run in' for a variable period of time before it will give a satisfactory result. As oxidation products accumulate the flow of the water will be progressively restricted. When the flow is no longer adequate the filter sand has to be replaced or cleaned. Either procedure is inherently unpleasant!

The settlement and tank filter systems lend themselves to hand pumped situations but maintenance can be a problem. This could be reduced significantly by utilising valveless self-washing filters. The disadvantage of these is that when used singly they do not produce service water after a wash cycle until the filtered water/backwash reservoir has filled. A comparatively simple modification could divert a small percentage of the filtered water to the reservoir with the balance being delivered to service. In this way all users of the pump contribute to the maintenance of the system.

Apart from open tank detention many systems utilise a closed detention vessel ahead of the filter (Figure 6.1). This has the advantage of operating at the system pressure thus eliminating the necessity of incorporating an additional pump to force water through the filter.

Figure 6.1 Typical 'triplex' filter installation for iron/manganese reduction with aeration vessel ahead of backwashing permanent media filter. (*Shakesby and Sons Ltd.*).

Even if most of the iron entering filters is in the ferrous form the system will usually work because most filter media contains a catalyst, commonly manganese dioxide based but not exclusively so, or becomes a catalyst with use as with the 'Open Tank Filter.' This accelerates the reaction between the ferrous iron and the oxidant (usually oxygen from the air) resulting in ferric iron being held on the surface of the catalyst. When using manganese dioxide containing products there is the possibility in some circumstances of increasing the level of manganese in the water.

Filter washing

The iron is usually held sufficiently strongly to permit rapid starting and stopping of the water flow without displacement of the ferric Iron, although such flow changes are not ideal. Periodically the accumulated iron has to be washed out of the filter. The majority of systems employ a multiport valve to carry out this automatically. The wash can be instigated by a rise in differential pressure across the filter, on a volume treated basis or at fixed time intervals. For the smallest supplies it is best to keep controls as simple as possible for maximum reliability and employ this arrangement.

Washing normally involves reversing the flow of water through the filter and expanding the filter media sufficiently to displace the iron. This is called backwashing but it is not unusual for systems to fail due to inadequate flow during the operation. Lack of backwash flow does initially not stop a system working, but causes progressive failure. The backwash flow can be many times as high as the service flow rate and it is vital to ensure adequate water is available to carry out the operation. Sometimes multiple small filters are installed, each one washing at a different time to ameliorate this potential problem. This of course increase the system cost and the amount of servicing required. Most systems use the untreated water for the operation but when significant quantities of ferric iron are present reliability can be improved by using filtered water.

Another failure mechanism can be caused by a process called mud balling. As the backwash commences the media bed expands and this can cause a heavily loaded layer at the top of the filter bed to break into comparatively large lumps which then sink through the rest of the fluidised media bed. At the bottom of the bed these lumps eventually form balls of media which have a negligible filtering ability. In time there can be insufficient active media for the system to work satisfactorily. Mudballing can be reduced by more frequent backwashing or minimising the bed expansion but using air to agitate the bed to displace the iron. This allows longer filtration runs and minimises water usage for the washing process.

Process refinements

A common characteristic of iron is that it can deposit on any surface it contacts once it is able to oxidise. The actual deposit it forms can act as a catalyst thus accelerating the rate of deposition. Components intended to aerate the water and provide detention to allow the oxidation process to take place can become severely clogged with deposits that restrict water flow. Devices installed to vent surplus air can become inoperative or leak. If the flow path incorporates a pump its performance is likely to be compromised. As the flow falls (which may not be apparent initially because it will often have been sized to provide a flow higher than the service flow for backwashing purposes), the backwash rate will become inadequate causing eventual system failure. The rate of flow will also be restricted by deposits forming in any control valves (usually multiport automatic units). Because of these inherent shortcomings it is not unusual to find that suppliers of backwashing filtration systems recommend servicing at six monthly intervals.

In the early 1970's Fred Smith, a chemist, employed at the time by Anglian Water in the UK, realised that reliability could be greatly improved and servicing reduced by incorporating the aeration and filtration in one closed vessel. His first designs used an unwieldy, tall vessel which required excessive headroom. He later employed a shorter vessel but increased the diameter of the upper section to hold a greater volume of air. On some systems an oil free air compressor was used to introduce air into the filter but on others he utilised an ingenious system. The automatic multiport valves he used for filter washing functions were derived from valves normally used for the control of ion exchange water softeners. These incorporated a small Bernoulli Pump (usually called an injector) to draw a brine solution into a softener in order to regenerate the ion exchange resin but he realised they were equally as effective at drawing air into the filter vessel. Many manufacturers supply similar systems now. Because the air is only drawn into the filter during a part of the washing sequence the capacity between wash cycles is limited by the available oxygen entrained in the vessel.

The design could reduce or eliminate the problems of oxidation products depositing prior to the filter but the capacity is limited, the risk of mud balling is still present and the backwash flow rate higher than the service rate although Fred used a very light filter media which reduced the difference.

The limit on capacity caused by the finite volume of oxygen available can be eliminated by continually introducing air to the filter during service flow. Surplus air has to be vented from the filter although it is not unknown for some to operate while 'percolating', the presence of the surplus air in the water after treatment is unlikely to be acceptable. Care has to be taken in the design of any venting device to ensure it is not adversely affected by the iron. With a little ingenuity it is possible to utilise the aeration device (usually a Bernoulli Pump) to provide an air/water mix to perform the backwashing operation. This can result in the backwash flow being lower than the service flow. It is essential that the air/water is distributed over the total area of the filter bed. Using industry standard distribution systems designed for water only will simply result in a 'boil' of air over a small area of the bed which can result in poor cleaning and rapid system failure. A properly designed distributor removes the need to incorporate an under bed. This allows the use of a deeper filter bed giving a longer contact time with its associated improvement in removal capacity. Servicing is also greatly simplified. An example of good and bad designs for iron and manganese reduction is shown in Figure 6.2.

Figure 6.2 Examples of good (left) and poor (right) installation and design. Process on left shows two stage system for iron and manganese reduction in dedicated plant room with room for service access. Example on right shows 'triplex' system with unnecessary break tank installed ahead of aeration and limited access for maintenance. (*Shakesby and Sons Ltd.*).

Problem waters

When aeration systems do not provide a satisfactory reduction of iron it is not unusual to find that the pH is just too low for the reactions to take place quickly enough if at all. Sometimes this is caused by carbon dioxide. A simple test to determine if this is the case is to cascade an aliquot of water, say a few hundred millilitres, between two containers 10–20 times and check the pH before and after. A rise in pH after indicates the presence of CO_2. It is possible to use de aerating towers or cascading systems to reduce CO_2 and raise the pH. Unfortunately this is likely to result in the deposition of iron oxidation products with their associated problems described earlier. The arrangement may be appropriate with 'Open Tank' filters where labour may be readily available to carry out regular cleaning operations.

It is possible to incorporate some sacrificial alkaline material in filter beds. These may be minerals composed primarily of calcium carbonate, magnesium oxide or combinations of both (magnesium dolomite). Some are processed natural products and some synthetic. Magnesium products have an inherently higher pH and can elevate the pH of some waters to above 10 although this is usually after the treatment plant has been quiescent for some time. Although this is above some national water quality standards, the effect is not likely to be harmful. It is generally considered that the magnesium products are more active than calcium products and tend to be more popular for pH elevation in iron reduction systems. Unfortunately the iron tends to deposit on the alkaline material and essentially shield the water from the active surface even when vigorous backwashing is employed. An alternative arrangement is to precede the iron filter with a pH elevating unit. This may consist of a vessel containing the active media where the water simply percolates upwards through the media or downwards when backwashing is usually employed periodically. Even with anaerobic waters a reaction can still occur between the media and the iron which isolates the water from the media.

It is possible to elevate pH by injecting an alkaline solution into the water. Solutions of sodium carbonate and sodium hydroxide are often mentioned as appropriate. Sodium carbonate is commonly used to elevate the pH in swimming pool water but it is necessary to make up a solution before it can be used in a water supply. Sodium hydroxide is available in various solid forms and also has to be made into a solution. It is highly exothermic and caustic causing serious burns on skin contact. Eye contact is likely to result in blindness. Extreme care has to be taken handling the chemical with full eye protection, preferably a full face mask, hand and arm protection. It is possible to obtain it in solution but will often require further dilution before it is suitable for use. Excessive dosing of the chemical could cause fatal results and great care has to be taken that any injection system is turned off if there is no water flow. To this end dosing systems which use the water flow to provide the energy to facilitate the injection are the safest. System design must also ensure that any chemical cannot siphon into the water or be drawn in if negative pressure conditions occur. Both sodium carbonate and sodium hydroxide will often require dilution with soft water if problems with precipitation products are to be avoided.

If pH is elevated to an excessively high level periodically, this can produce too rapid a reaction with the resultant formation of colloidal iron. This form of iron will pass through the following filter thus defeating the purpose of the pH elevation. Reduction in the dosing solution strength can ameliorate the problem but it must be remembered that a greater volume of alkali solution will be used for the same overall pH elevation.

When elevating pH it is prudent to remember that pH is a logarithmic scale and that to raise the pH by a factor of 1 requires a tenfold increase in hydroxyl ions and to elevate by a factor of 2 a one hundredfold increase and so on.

Other oxidation regimes

It is possible to use other oxidants such as ozone, chlorine, potassium permanganate, chlorine dioxide and hydrogen peroxide. Ozone has the advantage that it is generated electrically on site without the necessity of purchasing any chemicals but it is a toxic gas and care has been taken to ensure any surplus is appropriately disposed of. Chlorine is probably one of the most readily available chemicals in the form of sodium hypochlorite but care must be taken to ensure no harmful by-products are produced such as trihalomethanes, e.g. chloroform. Although chlorine dioxide is generated on site at least two chemicals have to be purchased and this is generally exclusively from the manufacturer of the generator.

All the aforementioned chemicals are prone to produce colloidal iron which can be difficult to remove. They may be used individually or in combination with oxygen from the air. It is generally not advisable to attempt to use more than one oxidant due partly to the risk of violent reactions between some of them. That said the use of chlorine injected prior to potassium permanganate ($KMnO_4$) is suggested by some as this arrangement has the potential of reducing the quantity of the more expensive oxidant.

Conducting pilot studies is even more important when using additional oxidants as there is the danger of making the perception of the iron worse, for example a water with a small amount of iron which is just about tolerable can be converted to an unacceptable yellow colour with chlorine addition. Any of the aforementioned may be utilised in conjunction with pH elevation.

Regenerating filter systems

The use of manganese greensand has been a fairly common method of iron reduction since the middle of the 20th century. It can be used in two processes. One is called Intermittent Regeneration (IR) and the other Continuous Regeneration (CR). The former is usually the easiest to manage with small supplies as the plant required is almost identical to the very common ion exchanger water softener. The regenerant is usually potassium permanganate but sodium hypochlorite can also be used. Care has to be taken that the materials used in the construction are suitable for continuous contact with the regenerant. As with many other types of filters the backwash flow rate can be far higher than the service flow rate. The volume of water treated between regenerations can be calculated and this information is very important because as soon as their capacity is exhausted they will pass untreated water. A metered system of regeneration initiation should be employed to prevent this.

There was a risk of damage to the active surface of the manganese greensand if overrun but the latest products do not actually contain any greensand and is manufactured using a different process resulting in a more robust product.

The CR method involves injecting a continuous feed of potassium permanganate or sodium hypochlorite ahead of the filter. Enough chemical has to be injected to oxidise the iron but not enough to result in the typical purple/pink colour of potassium permanganate in the treated water. When sodium hypochlorite is used, sufficient has to be injected to give a small chlorine residual in the treated water. The method involves the preparation of dosing solutions which need to be prepared with accuracy. There is the danger of hardness precipitation, especially around injection points when hypochlorite is used. For these and other reasons given earlier the IR system is generally the easiest to handle.

Regenerating ion exchange systems

Ion exchange water softeners usually work by exchanging the cations which cause hardness in water, calcium and magnesium, with sodium. They are also capable of exchanging iron, provided it is in the ionic form. As a result they elevate the sodium level in the treated water. If there is a requirement to supply unsoftened water for drinking this is not possible. They replace one type of cation with another and as such are not filters, which actually remove a substance.

Another problem with ion exchangers is that iron tends to oxidise in and on the exchange material irrespective of its chemical composition and block the ion exchange sites. As a result the capacity for both hardness and iron reduction falls. Some exchangers can be rejuvenated by using a chemical to remove the oxidised iron and sometimes this is fed in with the regenerant on a continuous basis. The method is inherently low cost but with high running costs in comparison to aerating system. It is a popular system as no great skill is required in installation provided a pressurised supply is already in situ. In spite of this long term results are often disappointing especially when manufacturer's parameters for satisfactory operation are ignored.

6.2 MANGANESE

Manganese is probably the second most problematic metal aesthetically. It is not generally considered harmful until levels rise above 400 micrograms per litre but this is subject to ongoing research and debate. The EU indicator concentration value is 50 µg/l but even this can cause staining in some circumstances, for example in swimming pools and dairies.

It is common to find manganese as well as iron in many waters with the manganese usually at an appreciably lower level. However there are situations where the iron is negligible and manganese concentrations of several thousands of microgrammes per litre occur. Because iron and manganese are so often found together they are usually considered as one subject when it comes to the treatment of small supplies, hence the term 'Iron and Manganese Filter'. When it comes to oxidation, the pH at which manganese oxidation occurs is theoretically, and often in practice, far higher than that for iron. If the pH is raised sufficiently to reduce manganese adequately it is likely that the iron will convert to a colloidal form which is difficult to remove without additional treatment. For municipal and many small supplies using a separate stage for manganese reduction can provide the best results and be simpler to manage (Figure 6.3).

Figure 6.3 Larger industrial system with two-stage bacteriological iron removal followed by manganese polishing filter (*Shakesby and Sons Ltd.*).

In spite of this it is not uncommon to find that the manganese concentrations are reduced significantly in an aerating iron filter where the pH is only around 7. In most situations provided there are a few thousand µg/l of iron present useful reduction will occur but it has been known for almost total manganese removal to occur with virtually no iron present. It is thought that the manganese may be adsorbed by the iron oxidation products, but this cannot explain the latter case. There are no hard and fast rules which will determine if an iron filter will reduce manganese to a desired level. It is essential to conduct a pilot trial in order to determine this.

Sometimes magnesium oxide based products are incorporated in the filter media to attempt to raise the pH sufficiently to achieve adequate manganese reduction. This can produce the desired effect, but it can be comparatively short lived due to the tendency of the alkali material to acquire an impervious coat of material which isolates the active material from the water. This results in frequent servicing but in spite of this, the savings achieved in some circumstances still make the system attractive. As mentioned in the preceding section on iron reduction it is also sometimes possible and appropriate to precede a filter with a vessel containing the alkali material.

If pH elevation is attempted by introducing an alkali solution such as sodium carbonate or sodium hydroxide, (noting the earlier cautions regarding the use of this chemical), it can be difficult, if not impossible, to adjust the dosing to give satisfactory removal of both metals. Using weaker dosing solutions can help but as with iron removal, the volume of solution will obviously increase. Too low a pH will result in poor manganese reduction and a pH high enough to correct this can cause the iron performance to deteriorate. The quantity of alkali used does not rise linearly with pH rise, for example if the pH has been raised to 7 using one unit of alkali to give adequate iron removal, to raise it to 8 would require an additional 10 units and to 9 an additional 100 units. The amount of pH elevation required has to be determined by on site tests. Careful attention must be given to the practicality, running costs and ultimate use of the water. With many agricultural uses the pH may need to be reduced if it has been raised to permit metal removal. This would involve the addition of an acid with its attendant complications and costs.

As with the processes used for the reduction of iron the same oxidants can be used if appropriate. Ozone can be effective but often at the expense of poorer iron performance, although it still may be adequate. The same can be said for all the alternative chemicals.

Often the simplest and safest way to reduce manganese to an acceptable/required level is to employ a manganese dioxide media (commonly manganese greensand) which is regenerated intermittently with potassium permanganate or sodium hypochlorite. The system is identical to that used for iron reduction with the capacity being calculated and regeneration instigated on a volume treated basis. When no appropriate regeneration chemicals are available it might be possible to use ozone which would be generated on site. Short term experiments using this method have yielded promising results. It is also possible to utilise a continuous feed of regenerant but adjusting this is more onerous than simply altering the pre set capacity of a treatment plant to accommodate changes in the untreated water. Another benefit of the system is that it will work satisfactorily at pH levels down to about 6 which render pH elevation unnecessary as far as the manganese reduction is concerned.

If iron is also present in the water there is always the risk that this will deposit on the surface of the manganese greensand and inhibit the manganese removal process. Unfortunately unless prolonged trials are conducted it will not be possible to determine if this will occur. Obviously if the iron is reduced to a low level in a preceding treatment stage this failure mechanism is unlikely to happen.

It is important to realise that the catalytic and other reactions that occur using different filter media normally take place on the surface of the material. Manganese dioxide (MnO_2) occurs naturally as an ore usually called Pyrolusite. It is sold under many trade names. The percentage of MnO_2 varies depending on the source but those with the highest level are likely to be the most effective. Pyrolusite has a very high specific gravity (SG) and hence requires a backwash flow rate typically several times as great as its service flow rate. By coating a material with a lower SG with MnO_2 the backwash flow rate can be reduced and a higher purity coating applied to the substrate. The only advantage of a material composed entirely of the solid oxide is that it has no coating to lose, but otherwise it will work no better. Modern coated materials can be very robust and their advantages usually outweigh any perceived shortcomings.

For those wishing to investigate iron and manganese systems in more detail a useful publication is '*Iron and Manganese Removal Handbook*' by Elmer O Sommerfeld, Published by the AWWA.

KEY POINTS

1. Iron is the main metal to affect most small supplies at concentrations requiring treatment;
2. Treatment for iron need not be complicated, however it does need to be appropriate to the specific situation;
3. Care should be taken to ensure that oxidation of ferrous iron is substantially complete prior to filtration;
4. Manganese dioxide may be used to successfully catalyse the deposition of ferrous iron on filter media in some cases;
5. Manganese removal processes rely on similar principles to iron, with catalytic removal using manganese dioxide probably the simplest;
6. A number of treatment processes for arsenic are available, with the most appropriate dependent upon circumstances of the supply.

REFERENCE

Sommerfield E. (1999). Iron and Manganese Removal Handbook. American Waterworks Association.

Chapter 7

Conditioning water to prevent dissolution of plumbing materials or post-treatment contamination

7.1 TREATMENT IS ONLY PART OF THE STORY

Consistent, effective treatment is vital to ensure that water is of satisfactory quality and safe and pleasant to drink. Even the most well-managed treatment process is of limited value, however, if the water is subsequently contaminated by the pipes and plumbing fittings between the treatment process and the location where the water is consumed. Microbiological contamination due to ingress or poor system design is a serious health risk, especially in supplies where there is no residual disinfectant or the disinfection process is not wholly robust, however this is beyond the scope of this publication.

Of possible concern is contamination of the water with metals that originate from corrosion of the plumbing system downstream of the treatment process. A number of metals can enter the water in this way, at concentrations that can cause aesthetic and health concerns. Additionally, trace amounts of metals in the water supply system, such as manganese, can accumulate in the corrosion deposits of premise plumbing and may subsequently slough off at greater concentration.

Significant concentrations of plumbing metals are often found in small water supplies for a number of reasons, primarily:

- use of inappropriate materials in pipes, tanks and plumbing fittings – for example, the use of copper pipes when the water has an acidic pH;
- lack of maintenance of plumbing systems;
- inadequate or no conditioning of water to make it less corrosive to metals.

Root causes are similar to those for the other water quality issues that affect small water supplies – lack of awareness and expertise amongst supply owners and users, lack of resources and limited regulation. Corrosion itself is an extremely complex subject that occurs in many different ways. The subject is simplified if it is considered in relation to potable cold water and domestic hot water only.

When a new municipal water source is being developed a number of parameters are usually considered in an attempt to ensure the water will not be plumbosolvent, dissolving any lead in plumbing systems. With small water supplies it is unusual for the same procedure to be followed. It is almost universally believed that a pH value below 7 will be aggressive towards metal fittings and pipe work and a value above 7 will not be. Although this is often the case, it must be understood that it is not always true and other parameters must be considered, just as with other aspects of water treatment. National standards may specify minimum and maximum levels of pH but the World Health Organisation does not make any recommendation, simply stating that the value is typically between 6.5 and 9.5 and that the most appropriate pH for a water depends on its other chemical constituents and the nature of any associated distribution network. Statements to the effect that a pH of 7 or above is essential or desirable and ones below this level will be aggressive to metal are an oversimplification of the science.

7.2 INDICATIONS AND EFFECTS OF POST-TREATMENT METAL CONTAMINATION IN SMALL WATER SUPPLIES

Where metals contamination is a problem in small water supplies, the source of contamination may not be immediately obvious, but it is important that the possibility of contamination post-treatment is considered. Although not fool-proof, useful indications that the source of a metals problem may lie within the distribution or plumbing system would be:

- Discoloured water and sediment;
- The presence of metals associated with plumbing materials (copper, lead, nickel);
- Staining of sanitary ware, often with green or blue deposits;
- Frequent bursts or leaks within the plumbing system indicating that corrosion is taking place;
- The presence of plumbing materials of old or unknown origin;
- The absence of pH correction or water conditioning where source water is known to be soft;
- Where the problem is thought to be new, any recent changes to the plumbing system, e.g. the installation of new pipes of fittings;
- Changes in treatment, particularly the introduction of oxidants, that can deposit metals from the water or mobilise them from existing corrosion scales within premise plumbing.

The presence of plumbing metals in water supplies can have both health and aesthetic implications. Aside from imparting a taste or visual issue to the water, there have been cases of dissolved metals staining hair. The adverse health impact of lead in water supplies, especially on the development of children, is well documented even if there is uncertainty as to a safe concentration. Current medical thinking is that there may be no concentration of lead that may be considered 'safe' and the metal's presence in water should be minimised as far as possible. Other heavy metals found in plumbing fittings, such as nickel, can cause adverse health effects, usually with long term exposure. In particular, exposure to nickel has been linked to dermatitis in those sensitised to it. Gastro-intestinal effects from high concentrations of metals have also been reported. In the United States, copper is regulated based on a combination of gastro-intestinal effects and Wilson's disease. Table 7.1 describes the metals most commonly found in water due to dissolution of plumbing materials and likely sources.

Lead and copper are probably the plumbing metals most commonly found at significant concentrations in samples taken from small water supplies and are worth considering in more detail.

Lead

Although lead piping has not been used for conveying water in new installations for many years and the use of lead containing solders is often prohibited there are still many situations where lead may be present. In such situations it can be found that excessive levels are found in the water at the point of use even when the pH is 7 or higher. This is not uncommon with soft waters and lead may still be detected at a pH above 8 or 9. Unless tests are conducted to determine if lead is present it will not normally be detected by the end user. The presence of lead piping, where it is visible is indicated by dull grey pipe, although the soft metal is shiny when cut. A very characteristic feature of lead piping is the rounded, bulbous joints.

Table 7.1 Metal sources in premise plumbing.

Metal	Possible Sources	Effects
Lead	Lead piping and tanks; solder; some brass fittings; stabiliser in some PVC pipes	Developmental issues in children; raised blood pressure in older people.
Nickel	Plumbing fittings made from metal alloys, stainless steel and plated products	Chronic exposure may cause heart and liver damage. Dermatitis in those sensitised.
Copper	Piping, brass fittings and fixtures.	Taste, discolouration of fittings and hair, possibly gastro-intestinal effects. Toxic to people with Wilson's disease.
Zinc	Galvanised piping and tanks, brass fittings	Taste and potential gastro-intestinal effects at high concentrations.
Iron	Iron or steel piping or tanks	Primarily aesthetic – sediment and brown discolouration. Substrate for accumulating contaminants from upstream sources.

Small amounts of lead in piping, fittings or solder can yield surprisingly high concentrations in water samples, especially where conditions favour plumbosolvency. Over time, pipe interiors can become coated with lead containing solids which offers a degree of protection from further corrosion of the lead itself. If cut or disturbed in anyway, or if water quality changes significantly to alter the chemical equilibrium, greatly enhanced lead concentrations, including lead particulates, can result.

It is difficult to reduce the plumbosolvent tendencies of water in small supplies – elevating pH may help but provides no guarantee; the presence of dissolved organic matter will complicate things and artificial protection methods such as dosing phosphate compounds are unlikely to be an option on grounds of both cost and practicality. Removal of all lead-containing components, and vigilance to ensure no new lead-containing fittings are inadvertently introduced, is the only lasting solution and is the approach recommended for small water supplies. In the short term, as a temporary protective measure, a substantial reduction in lead exposure can be obtained by ensuring water is only consumed after thorough flushing, particularly where water has been standing in pipes overnight.

Copper

This metal is widely used in internal plumbing systems, but corrosion problems with municipal water supplies are very rare, unlike the situation that occurs with many small supplies. For this reason, when installing water treatment plant on small water supplies it is often better to avoid using copper pipework. It will be beneficial to undertake a brief risk assessment prior to any installation of copper pipework, which at the minimum should focus on water quality, particularly pH and water softness/hardness; likely organic quality should also be assessed by reference to source water type. Sequential sampling (e.g. 12×1 litre samples) after a period of stagnation may reveal copper leaching problems but the costs involved will likely be prohibitive for small supplies. Regular flushing of copper pipework after its installation should minimise copper leaching problems.

Copper often manifests its presence by causing blue/green staining on sanitary ware especially where taps (faucets) or showers drip. Quite dramatic colour effects can occur with hair especially that which has been dyed or bleached. It is not unusual for new copper pipework to release high levels of copper into the water until a protective layer of carbonates and/or copper oxide form. Some waters are inherently cuprosolvent in which case no such layer forms. These waters are typically those with low alkalinity and low pH. Sometimes water can develop a distinctive blue colour, caused by Cu^{++} ions which do not precipitate out to form a protective layer.

The eventual result of copper corrosion will be leaking pipework. Pipework can thin until it perforates. This commonly occurs at bends and points where high velocities may occur. Some corrosion can be caused by carbon deposits in the pipework as a result of the manufacturing process but for this to occur the water has to be typically hard and with a pH above 7, which is contrary to often held beliefs. Copper corrosion is complex, and for a detailed examination of the types and causes of copper corrosion in plumbing systems the reader should refer to Oliphant (2010) – see references at the end of this chapter.

7.3 ESTABLISHING THE SOURCE OF THE PROBLEM

7.3.1 Investigative sampling

With new sources, especially if no established distribution system is in place it is possible to run the water through a small section of the proposed pipework during the planning period and check for any adverse effects. A decision could then be taken to use a different distribution material (e.g. Plastic.) if necessary.

If a small water supply is already in use, investigating any problems with metals will require a systematic approach:

- A series of 'spot' samples can be taken to check out metal contamination of source waters. The number of samples and the sampling period will depend on the variability of contamination. Generally, groundwater quality is fairly stable, springs less so, and the most variable sources will be surface derived (streams, lakes, etc). The commonest problems are with aluminium, arsenic, iron and manganese.
- If the water is treated, then both untreated and treated water should be sampled at the same time. Most treatment processes should reduce metal contamination, but if coagulant chemicals are used contamination with iron or aluminium is very possible.
- The distribution of treated water to the point(s) of use will normally require pipework, valves and possible a storage tank. These should first be inspected before attempting to sample. Sampling points can then be selected that match the circumstances of the distribution arrangements. 'Spot' sampling over a period of time at a sequence of sampling points should provide an indication of any problems with metals, the likely focus being iron from iron pipes and tanks, and lead from brass components (meters, stop-cocks, etc). To further investigate iron problems from iron (or galvanised iron) pipes, sampling during and after flushing will indicate if rust solids have built up in the pipes.
- Metal pick-up from premise plumbing is more difficult to investigate because metal concentrations depend on how long the water has been in contact with pipework, fixtures and fittings. The commonest problems are lead from lead pipes and brass fittings, copper from copper pipes and nickel from chrome-nickel plate components (particularly taps). Sequential sampling (e.g. 12×1 litre samples) after a period of stagnation should reveal the extent of any metal leaching problems.

From the above, it can be appreciated that investigative sampling may require a significant amount of sampling and analysis, and this may not always be feasible when resources are limited. At its simplest, an occasional check sample of the water at its point of use should be obtained, combined with visual assessment (staining, discolouration) and a knowledge of the water supply system.

7.3.2 Indices

There are a number of corrosion indices that have been purported to predict whether the characteristics of the water are pre-disposed to the dissolution of plumbing metals, but these are generally unsuccessful for copper and lead. A detailed examination of indices is outside the scope of this publication, but by far the one most commonly referenced to is the Langelier Saturation Index.

It is often recommended that to determine if a water will be corrosive towards a metal that the Langelier Saturation Index (LSI) must be determined. It is important to understand that the index was developed to determine whether water would deposit a carbonate film on surfaces or dissolve it, or in other words be aggressive to a carbonate deposit, not a metal. It is particularly useful to define the limits to pH adjustment without the necessity for softening or bicarbonate removal, and it could have some value for predicting the corrosivity tendency of some waters towards materials whose performance is improved by the integration of hardness and carbonate ions into the pipe scales. Unfortunately, it is highly unsuitable for predicting lead or copper release, and cannot be accurately computed in the presence of significant amounts of natural organic matter or calcium carbonate nucleation inhibitors like various phosphates.

The LSI is the difference between the measured pH and the value of the pH calculated from the temperature, the ionic strength, the calcium hardness and the total alkalinity. A positive value indicates that the water should deposit calcium carbonate i.e. be scale forming. A negative value indicates it is not scale forming thus aggressive towards calcium carbonate. In practice this means that a soft water with low alkalinity and high pH could be corrosive to iron or cementitious materials, even though the LSI may be positive. Another contradictory situation can occur if there are other ions such as chloride or sulphate in the water that may inhibit scale formation, or cause elevated scale porosity and facilitate increased corrosion rates of the metal oxide/carbonate/hydroxide scale.

7.4 FACTORS CONTROLLING THE CORROSION OF METALS INTO SMALL WATER SUPPLIES

Where the relevant materials are present in a plumbing system, the extent to which they corrode and dissolve into the water supply, and therefore the concentration of metal in the water, will depend on a number of properties and variables. These properties, and the mechanisms controlling them, are often complex and beyond the scope of this publication, however a number or references have been provided that give further information.

Contact time

Where water is allowed to stagnate in contact with a source of metal, the concentration will increase with time, particularly in the presence of oxidants, such as chlorine or oxygen. Concentrations tend to increase in proportion to stagnation time until an equilibrium concentration is reached with the stable form of the metal under the prevailing redox conditions.

Temperature

In general, increased temperature speeds up chemical reactions and dissolution of plumbing metals will accelerate where water temperatures are higher, though there are exceptions when corrosion inhibiting scale solids decrease in solubility with temperature, as some oxides and carbonates do. Where plumbing systems are poorly designed, and cold water systems run close to warm water ones, or other sources of heat, this can greatly increase the extent of any problem. Temperature changes in source waters may also coincide with seasonal changes in water chemistry, which may combine to change corrosivity and metal release into the water.

Galvanic effects

Where plumbing systems use more than one type of metal in contact with each other, the difference in electropotential between the different metals can cause galvanic corrosion where the rate at which metal ions are released into the water is greatly increased. Galvanic corrosion is affected by the background water chemistry, particularly parameters such as pH, amount and type of disinfectant, chloride, sulphate and alkalinity. So, the extent to which the same metal connection will or will not cause a problem may change from system to system because of different water qualities.

Age of system

As plumbing systems age, their tendency to corrode and dissolve metals into the water supply may change. Leaching of metals from a newly installed system may settle with time to acceptable levels. In other circumstances, aging pipework may corrode at an increasing rate and cause problems with discoloured water, sediment and non-compliance with regulatory water quality standards.

pH and alkalinity

Typically, the solubility of corrosion product increases as water pH decreases, however this is an over-simplification of a complex subject. It is true that many of high metals concentrations in small water supplies are due to low pH, low alkalinity waters – too often water conditioning and pH correction are afterthoughts or seen as 'optional extras'. Certain types of copper corrosion, sometimes responsible for blue water colouration or pinhole leaks, occur at high pH and low alkalinity. There have been cases of this occurring on small water supplies where inappropriately sized pH correction filters have been installed, resulting in excessive contact times with filter media and high pH.

Oxidising chemicals

The presence of oxidising chemicals in the water can change its properties and alter the rate of corrosion and the solubility of corrosion byproducts and pipe surface coatings. This would generally be assumed to be an increase, and evidence suggests that this is more prevalent with iron and copper, but less so with lead. Having said this, the disinfection of water is paramount to safeguarding health, and corrosion issues should be resolved without compromising microbiological safety.

Natural organic matter (NOM)

The presence of naturally occurring organic matter in the water can have an impact on the corrosive properties, although it is often not considered. The effects are not always well studied and can be different depending on the metals concerned. The main direct effects of NOM are due to the formation of complexes with metal ions. NOM has been found to increase the amount of soluble lead in water, and this appears to be especially significant in low pH, low alkalinity waters. It is apparent that it is desirable to reduce the concentration of NOM in waters in order to minimise this effect. This can be difficult to do effectively on small water supplies, but it does bring other benefits such as the reduction of fouling and biofilm growth, enhancing disinfection and oxidation-reduction potential stability, and reduction in disinfection by-product formation.

Chloride and sulphate

The concentration of these anions in the water can have a significant impact on certain types of corrosion. The effect depends on the metal, the concentration of the anion and their ratio to one another.

7.5 THE CONDITIONING OF WATER TO MINIMISE CORROSION

If treatment is required it is advisable to consider all the parameters that could influence the method used, although initially in practice only pH is usually considered. The simplest method of elevating pH is to pass the water through a bed of alkaline granular material. The alkaline material used is usually calcium carbonate or magnesium oxide. There is a naturally occurring compound of both of these; magnesium dolomite. The materials are processed to render them more suitable for water treatment use.

If the water is free of particulate matter it is not unusual to pass water upwards through the media. The contact time to achieve pH equilibrium can be prohibitively long but times of a few minutes often have the desired effect. Problems can occur due to the fact that apart from elevating the pH the hardness is also increased. This can give rise to complaints from users especially if they are used to very soft water. Where there is doubt about the suitability of utilising an upflow system it is prudent to utilise a downflow arrangement with a backwashing system to clean the bed. The flowrate required for this to ensure adequate bed expansion is usually appreciably higher than the service flowrate and care must be taken to ensure an adequate flow of water is available. Even when backwashing systems are employed the alkaline material can agglomerate into large accretions which have low activity and make remediation difficult. Waters with quite small levels of phosphates are particularly prone to cause this problem. It is also possible for carbonates to deposit on the media and produce a similar effect as phosphates.

Another problem is that the water will often become scale forming and if ultraviolet disinfection is installed downstream its efficacy can be rapidly diminished, with obvious implications for microbiological safety and requirements for maintenance. This effect is not surprising as the purpose of the treatment is often to increase the LSI. It may be considered appropriate to install any pH correction downstream of a UV filter, however the potential for microbiological growth on the filter media and pipework should be considered.

Water containing organic compounds, notably humic and fulvic ones which are typical of upland peaty water sources can deposit a coating that effectively isolates the water from the alkaline material thus rendering the treatment ineffective.

It is possible to elevate the pH by injecting an alkaline solution of sodium carbonate or sodium hydroxide. This has serious health and safety implications that may be difficult to adequately manage on a small supply system run by non-professionals. Their use does not increase the hardness of the water but more intervention is inherently involved when dosing is used and when considered with the health and safety issues involved in handling may render this approach impractical for most small water supplies managed by non-professionals. An example of a neutralizing filter arrangement is shown in Figure 7.1.

Figure 7.1 Dual neutralising filters installed on a small water supply.

KEY POINTS

1. Significant concentrations of metals can be achieved through dissolution of corrosion products of metals used in plumbing fittings and pipework.
2. Treatment to condition water to reduce corrosivity is key, yet often neglected in small water supplies.
3. Indices have limited application in gauging the corrosivity of the water.
4. A number of factors in combination control the rate at which corrosion occurs.
5. Plumbing materials should be chosen carefully to reduce the risk of adverse effects due to corrosion.
6. Simple filters containing alkaline media may be used to increase pH and alkalinity on small water supplies.

FURTHER READING

DWI/DEFRA (1991). Deterioration of water quality in distribution systems – the effects of water quality arising from in situ cement lining, WRC Swindon, UK, Report DoE 2435-SW(P).

Health Canada (2009). Guidance on Controlling Corrosion in Drinking Water Distribution Systems. Water, Air and Climate Change Bureau, Healthy Environments and Consumer Safety Branch, Health Canada, Ottawa, Ontario (Catalogue No. H128-1/09-595E).

Korshin G. V. and Ferguson J. F. (1999). Corrosion and Metal Release for Lead Containing Materials: Influence of NOM, University of Washington.

Moore M. (2001). Chlorine's Effect on Corrosion in Drinking Water Systems. National Drinking Water Clearinghouse, West Virginia University, 2001.

Oliphant R. J. (2010). Causes of Copper Corrosion in Plumbing Systems, Foundation For Water Research Report FR/R0007.

Peng C. Y., Korshin G. V. and Ferguson J. F. (2013). Effects of Chloride, Sulphate, and Natural Organic Matter on the Accumulation and release of Trace level Inorganic Contaminants from Corroding Iron. *Water Research* **47**(14), 5257–5269.

Chapter 8

Manual of individual metals – occurrence, implications and management

ABOUT THIS CHAPTER

This section is designed to provide an easy to use point of reference for a wide range of metals, some of which are common in small water supplies, others being highly unusual. Metals are listed in alphabetical order.

Likely sources of the metal are given, and when faced with a potential problem these should be used in the first instance to determine the likelihood of the metal actually being present in the supply, or of some form of artificial indication, perhaps due to sample contamination or analysis error.

A summary of health and aesthetic implications is given – the definitive reference here is the WHO Guidelines for Drinking Water Quality, and this may be consulted for further, more detailed information. Common regulatory standards in use around the world are shown, where available. The European and US standards have been taken as a starting point, and local requirements should be checked.

Options for management of the metal in the catchment, to avoid the need for treatment, are provided where applicable. In many cases the only feasible approach is to seek an alternative source, where one is available. It is advisable to consider the option of control in the catchment when seeking to manage a metals problem, as this is almost always preferable to installing complex and expensive treatment. Even where catchment management cannot reduce the concentration of the metal to acceptable levels it may be sufficient to reduce the complexity of the treatment required or extend the life of treatment consumables. Thought should be given to the reliability of any catchment measure, and continued monitoring for the metal is advised, especially where the metal is harmful or increased concentrations are not readily noticed in the water due to aesthetic effects.

Options for treatment at source and at point of use are given. The former relates to the treatment of the whole supply prior to distribution to the point at which it is consumed, perhaps serving a number of properties. Point of use treatment relates to small scale processes available to treat the water at a single point of consumption. If the latter are used it should be remembered that water quality may vary considerably between different taps, even in the same property.

Treatment processes suggested are based on published information – references are at the end of the section. The efficacy of a process will affected by a number of variables and site-specific conditions. In all cases it is advisable to thoroughly investigate the characteristics of the water supply and wider circumstances prior to selecting a treatment process, seeking specialist advice as necessary.

ALUMINIUM

Occurrence	Aluminium is a metal element, abundant throughout the Earth's crust. In small water supplies it can occur naturally or be introduced as part of the treatment process. Aluminium salts are widely used in water treatment as coagulants to reduce organic matter, colour, turbidity and micro-organism levels, although flocculation is a comparatively unusual treatment process to be found in private water supplies (it is, however, common on larger public ones). Naturally occurring aluminium can also be present in private water supplies, at concentrations in excess of the regulatory standard. This is usually in a combined form, either as an inorganic salt or complexed with organic material such as humic acids. Aluminium is more likely to be leached from soils where water pH is low. Drinking water comprises only a small proportion of the route by which humans ingest aluminium, with food being the major source.
Chemistry	Aluminium has atomic number 13 and is commonly found in oxidation states +2 and +3. A number of aluminium compounds are readily soluble in water, with solubility increasing at lower pHs. At higher pH, aluminium may be found as a hydroxide.
Health and Aesthetic Significance	Aluminium is abundant in a range of foodstuffs, and drinking water is likely to be only one of a number of ingestion routes. A potential link between aluminium and Alzheimer's disease has long been suggested but not conclusively proven. Elevated concentrations of aluminium in drinking water can adversely affect those on kidney dialysis. Aesthetic effects of aluminium are primarily discolouration, often in a pale gelatinous form, although high concentrations can also impart a bitter or metallic taste to the water.
Regulatory Standards and Guideline Values	WHO do not set a health based guideline value but do suggest that residual aluminium concentrations should be minimised with 100 microgrammes per litre being achievable in larger supplies and 200 microgrammes per litre in smaller ones. Both USEPA and the EU set standards of 200 microgrammes per litre on aesthetic grounds.
Options for Management in Catchment	Unless there is an obvious point source of industrial origin, it is likely to prove difficult to manage aluminium that is derived from catchments. Catchment management to avoid the circumstances that can result in mobilisation of naturally occurring aluminium from soils due to activities such as de-forestation may be worth considering in some catchments. It is not unusual to find elevated levels of aluminium in newly drilled boreholes and dug wells. With continued abstraction of water this will usually drop to an acceptable level.
Options for Treatment at Source	Although less common in small supplies, poorly optimised treatment process where aluminium is used as a flocculant can result in significant concentrations of aluminium in the water, and such issues should be addressed before seeking any other solution. Depending on the characteristics of the supply, it may be practical to consider adsorption of dissolved aluminium onto a filter media. Various media may be effective, including granular activated carbon. Cationic ion exchange is effective, but regeneration of the ion exchange resin can prove difficult and require periodic treatment with acid, which is beyond the scope of an untrained water supply operator.
Options for Treatment at Point of Use	Small scale point of use RO units are the preferred mechanism for removing aluminium.

ANTIMONY

Occurrence	Antimony is a metalloid element, whose commonest natural ore is Stibnite. China is the largest commercial producer of antimony, although deposits are known elsewhere. Industrially, main uses of antimony include fire retardant chemicals and as a replacement for lead in some soldering alloys which may be used in plumbing – this may be one of the main routes by which it can enter drinking water. Additionally, there has been evidence that detectable quantities of antimony may leach from PET plastic bottles, and it would be reasonable to consider contamination of samples via this route at an early stage should antimony apparently be detected in a small water supply.
Chemistry	Metallic antimony is a hard, brittle silvery substance. It has atomic number 51 and, similarly to arsenic with which is shares a number of characteristics, is most commonly found in water with oxidation states +3 and +5. Although native antimony is insoluble, a number of antimony compounds are readily soluble in water.
Health and Aesthetic Significance	The toxicity of antimony depends on the form in which it is found. The soluble form antimony (III) is generally considered to be the most toxic. Most compounds of antimony do not have an aesthetic effect on drinking water and are therefore not usually detected by the consumer.
Regulatory Standards and Guideline Values	There is a WHO Guideline Value of 20 microgrammes per litre in drinking water. The USEPA Maximum Contaminant Limit is 6 microgrammes per litre. The EU Drinking Water Directive sets a standard of 5 microgrammes per litre.
Options for Management in Catchment	Where there is an obvious point source of industrial origin, it may be possible to reduce the concentration of antimony in the water by identifying and removing the contamination. If contamination is more widespread or the antimony is naturally occurring, this may be less feasible and the only option may be to locate an alternative source.
Options for Treatment at Source	Prior to considering treatment for antimony, more obvious routes by which it might enter drinking water should be investigated and addressed – these being leaching from plumbing fittings and unrepresentative sample date caused by contamination from inappropriate plastic sample bottles. Although coagulation and filtration is likely to be an effective treatment, this is impractical for most small supplies. Although the chemistry of antimony is complex it is likely that most forms will bind to aluminium or iron flocs that form in conventional coagulation processes and will, accordingly, be removed as part of the process. Reverse osmosis membrane (RO) provides reliable removal with minimal use of chemicals. However, issues such as fouling of the membrane and disposal of the waste stream may need to be considered. There is some evidence that adsorption onto ferric oxide media may provide effective removal, in much the same way as this has been used to treat arsenic. This may provide an economic and relatively simple means of treating antimony in small supplies.
Options for Treatment at Point of Use	Small scale point of use RO units or, potentially absorption onto metal oxide media.

ARSENIC

Occurrence	Arsenic is a metalloid element that may be found across the world at concentrations of health significance in waters used for drinking water. Arsenic in drinking water is a major problem in some parts of Asia such as Bangladesh, where it is having devastating health effects on communities. Arsenic is commonly found as sulphide minerals, and may also be associated with iron and manganese deposits. It is easily solubilized in groundwater, and both natural and man-made sources of contamination are not uncommon. Arsenic has many industrial uses, including wood preservatives, glass and semiconductor manufacturing. Concentrations of up to 5 mg/l of naturally occurring arsenic have been detected in groundwaters, with concentrations of 400 mg/l being detected in waters influenced by industrial pollution and mine drainage.
Chemistry	Elemental arsenic usually has the appearance of a shiny grey metallic solid. It is usually found in nature as a compound with other elements, primarily sulphur and iron. It has atomic number 33 and is most commonly found in water with oxidation states +3 and +5, although other oxidation states exist. Arsenic compounds can be soluble over a wide range of pHs and redox conditions, which partly explains why arsenic can be so problematic in drinking waters.
Health and Aesthetic Significance	Arsenic's toxic properties are well known and it is one of WHO's 10 chemicals of major public health concern. Arsenic is both acutely and chronically toxic to humans. The extent of toxicity is dependent on the chemical form of the arsenic. Acute symptoms of arsenic poisoning include vomiting, abdominal pains and diarrhoea. Prolonged exposure to arsenic can cause skin lesions, skin, bladder and lung cancers. There is some uncertainty over the toxicity of arsenic at low concentrations. The presence of arsenic in drinking water does not cause a change in the aesthetic nature of the water, so it is usually undetectable unless associated with other contamination.
Regulatory Standards and Guideline Values	There is a WHO Guideline Value of 0.01 mg/l in drinking water. USEPA Maximum Contaminant Limit of 0.01 mg/l. The EU Drinking Water Directive also sets a standard of 0.01 mg/l.
Options for Management in Catchment	These may be limited. Another source of water, where available, may be the simplest option.
Options for Treatment at Source	Coagulation is a conventional treatment process that can be used to remove arsenic from water using chemicals such as aluminium sulphate, and ferric salts, such as ferric chloride or ferric sulphate. Although traditionally employed on larger water supplies, the process could be adapted for use on smaller supplies where sufficient space and expertise existed. Coagulation of arsenic transforms dissolved arsenic into insoluble compounds such as aluminium hydroxide which are precipitated in a sedimentation

tank or separated by means of sand filtration. Chemical precipitation processes usually require arsenic to be in the form of As(V); if arsenite, As(III), is present, it is usually pre-oxidized using chlorine, permanganate, or ozone. This process requires regular monitoring and adjustment and the usage of potentially hazardous chemicals, making it less appropriate for many small water supplies.

Arsenic often occurs with iron and provided adequate iron is present the arsenic will often reduce to below the desired value in an efficiently functioning iron filtration system. It appears that, in these systems, oxidation occurs with oxygen alone, although this is usually a very slow reaction. Where the iron level is inadequate, including waters with no iron present, a solution of an iron salt such as ferrous sulphate can be injected. This has the benefit of being a low cost and generally readily available chemical but care must be taken that it is a grade appropriate for water treatment use.

Adsorption processes are perhaps more useful for typical small water supplies. Typical adsorbents for arsenic removal are activated alumina, activated carbon, granular ferric hydroxide, zeolites, manganese greensand, manganese dioxide, hydrous iron oxide particles and iron oxide-coated sand (Crittenden et al., 2005; Mohan & Pittmann, 2007; Collivignarelli et al., 2011; Camacho et al., 2011; Giles et al., 2011; Mondal et al., 2013).

Activated alumina allows removal yields of 95% both for As (III) and As (V), even if its treatment capacity depends upon pH, initial arsenic concentration, speciation (Kartinen & Martin, 1995; USEPA, 2000; Singh & Pant, 2004) and competing ions in the water. This material can be regenerated but both sodium hydroxide and sulphuric acid are usually employed. The number of regeneration cycles may be fairly limited and the spent regenerants may contain levels of arsenic that make their disposal difficult. The material may also be used on a one pass basis where it is readily available. The disposal of the spent material is not normally as difficult as with the regenerants.

Ion exchange (usually anionic) can enable arsenic removal yields higher than 90% and will typically treat 250–1000 bed volumes. If the total dissolved solid concentrations in raw water is higher than 500 mg/l, this process may be less effective.

The most suitable membrane filtration processes for arsenic removal are nanofiltration (NF) and reverse osmosis (RO). Using these types of membrane, arsenic can be removed both in organic and inorganic form, as long as it is in the oxidized form. Using these membranes the As(V) removal yield is higher than the As(III) removal yield (Conio & Porro, 2004).

Examples of treatment processes used for arsenic removal on small water supplies may be found in Section 10.1.

Options for Treatment at Point of Use

The adsorption, ion exchange and membrane technologies described above are most readily scaled to point of use applications. Adsorptive technologies are likely to be the most straightforward, although careful monitoring is required to ensure adsorption filters are changed when media is approaching exhaustion and arsenic concentrations begin to increase.

BARIUM

Occurrence	Barium in drinking water primarily comes from natural sources as barium compounds are relatively common. Many igneous rocks and some sedimentary rocks contain barium at some concentration. There are many industrial uses, all of which could be potential sources of man-made contamination. These include manufacture of rubber, plastics, electronics, textiles, glass, ceramics and bricks. Drilling wastes from the oil and metals industries often contain barium, and these can be the primary sources of man-made contamination of drinking waters.
Chemistry	Barium is a soft, silvery white metal in its elemental form, although it is always found as a compound with other elements such as sulfates and carbonates. It has atomic number 56 and exists in an oxidation state of +2. The solubility of compounds varies, but nitrate and halide (fluoride, chloride, bromide) salts are the most soluble. The solubility of barium compounds increases as water pH decreases.
Health and Aesthetic Significance	Acute effects of ingestion of very high barium concentrations are primarily gastrointestinal. Probably of more concern for long term exposure to levels of barium more commonly found in drinking water in excess of health guidelines is a potential link to hypertension. Barium salts do not impart a taste or odour.
Regulatory Standards and Guideline Values	There is a WHO Guideline Value of 0.7 mg/l in drinking water. USEPA Maximum Containment Limit of 2 mg/l. The EU Drinking Water Directive does not set a standard for barium.
Options for Management in Catchment	In some areas where barium is naturally present due to geological conditions, concentrations are likely to be fairly uniform in groundwater across that aquifer, meaning that drilling an alternative groundwater source nearby is unlikely to be effective in reducing levels. Surface water sources in an area are also likely to contain fairly uniform concentrations, however this can only be determined via sampling. Where the source is man-made and can be located, it may be possible to remove the contamination or divert the most contaminated water away from the point of abstraction.
Options for Treatment at Source	Lime softening can be used to precipitate barium along with other hardness minerals, although this is also a specialist technique probably best suited to larger supplies.
Options for Treatment at Point of Use	Cationic ion exchange is a common and effective option for barium removal on small supplies. As with all treatment processes, some maintenance is required including restocking of the salt used to regenerate the ion exchange resin. Reverse osmosis membranes will also be effective, although may be costly and require disposal of concentrate. It is known that some precipative softening techniques can reduce concentrations of the substance, but these are difficult to manage with small supplies. Ion exchange softening will remove barium along with other cations such as calcium and magnesium. When operated in the sodium cycle there will be an elevation of sodium in the treated water which may be undesirable.

BERYLLIUM

Occurrence	Beryllium is uncommonly found at concentrations of health concern. Where it does occur, beryllium may be naturally occurring or derived from man-made sources. In nature it is associated with some geological formations including certain types of granite. Primary manmade sources include combustion of coal and discharges from metal processing, electronics and defence industries. High concentrations of beryllium in water are very unusual, but tend to be due to point source contamination.
Chemistry	Beryllium is an alkaline earth metal with atomic number 4. Most compounds of beryllium are insoluble in water.
Health and Aesthetic Significance	While beryllium is harmful via inhalation, and these effects are well documented, there is less evidence for the health effects of ingestion. One study, involving dogs, did suggest gastrointestinal effects from ingesting high concentrations.
Regulatory Standards and Guideline Values	WHO does not set a provisional guideline value for beryllium on the basis that it is unlikely to be present in drinking water at concentrations harmful to health. It does tentatively suggest that, were it required, current evidence would point towards a 12 microgrammes per litre health based value. The USEPA sets an MCL standard of 4 microgrammes per litre for beryllium. The EU currently does not set a standard for beryllium in drinking water.
Options for Management in Catchment	Where beryllium occurs due to natural geology it is likely to be ubiquitous across the area and an alternative type of water source should be sought. Point source contamination due to industry such as the combustion of coal products could be remediated or an alternative, less contaminated source sought.
Options for Treatment at Source	Reverse osmosis, adsorption by activated alumina filter media and, potentially, ion exchange could be used for removal of beryllium at source or point of entry. Conventional water treatment processes, such as coagulation, clarification and lime softening may be effective at removing beryllium to concentrations below those considered potentially harmful, although these treatment processes are less used on small water supplies.
Options for Treatment at Point of Use	Point of Use reverse osmosis and activated alumina filters are probably the most feasible removal options. The treated water needs to be carefully monitored in order to determine the capacity of a system. It must be remembered that other constituents of the water can compete for adsorption sites on the alumina.

BORON

Occurrence	Boron occurs naturally in the environment, especially in association with geological deposits associated with marine sediments and geothermal fluids. It is especially common in arid regions – the Western USA and Turkey are the largest producers of boron minerals. Many igneous rocks and some sedimentary rocks contain barium at some concentration. Parts of the Mediterranean basin have significant concentrations of boron in drinking water and problems with elevated concentrations in drinking water have been recorded in Turkey and other Mediterranean countries such as Cyprus, Greece, Italy and Spain, with 14% of samples being found to be above the standard in the EU Drinking Water Directive of 1 mg/l. There are many industrial uses, all of which could be potential sources of man-made contamination. These include manufacture of steel, glass, fertilisers, pesticides and as cleansing agents.
Chemistry	Boron is a grey-black non-metal in its native form, although it usually occurs in combination with other elements, often oxygen. It has atomic number 5 and exists in an oxidation state of +3. Boron compounds have varying solubility.
Health and Aesthetic Significance	Acute effects of ingestion of very high concentrations of boron compounds may cause nausea and vomiting and skin irritation. There is some evidence that boron may have some impact on male reproduction. Boron salts do not impart a taste or odour.
Regulatory Standards and Guideline Values	There is a WHO Guideline Value of 2.4 mg/l in drinking water. The USEPA does not set an MCL for boron, but has set health advisories with a reference dose (RfD) of 0.2 mg/kg/day The EU Drinking Water Directive sets a standard for boron of 1 mg/l.
Options for Management in Catchment	Options of the management of boron in the catchment are limited, with the location of an alternative source often the most effective option. Boron's association with arid regions often makes this impractical.
Options for Treatment at Source	Treatment options are limited. Conventional treatment technologies such as coagulation are not effective for boron and are impractical for smaller supplies in any case. Specialist ion exchange resins targeted to boron removal are most likely to prove effective.
Options for Treatment at Point of Use	Boron-specific ion exchange is the most promising technology for point of use treatment, however whether this is scaleable to point of use applications is not clear. The process involves the use of hydrochloric or sulphuric acid and sodium hydroxide with the attendant safety issues. Rather than attempting treatment, an alternative water source should be sought if possible. There is some evidence that reverse osmosis membranes are not entirely effective for boron removal.

CADMIUM

Occurrence	The main mineral of cadmium is greenockite, and is commonly associated with zinc, lead-zinc, or copper lead zinc ores. Asia is the primary cadmium-producing region in the world, especially China, South Korea, and Japan. Cadmium and its salts are widely used in industrial activities, including alloys, nuclear reactor control rods (neutron absorption), paint, plastic stabilizer, fluorescent powder, insecticides, fungicides, electroplating, and batteries. Contamination of rivers and sediments in industrial areas is not uncommon, and there have been a number of high profile cases around the world, with concentrations many times the safe limits.
Chemistry	Cadmium is a soft, malleable, ductile, bluish-white metal. It is similar in many respects to zinc but forms complex compounds. Although cadmium usually has an oxidation state of +2, it also exists in the +1 state. Cadmium compounds have varying solubility.
Health and Aesthetic Significance	Food and water is the main source of exposure to cadmium. Cadmium is one of the most toxic metals in both natural and engineered waters. Cadmium ions in water do not impart a discernible taste or odour.
Regulatory Standards and Guideline Values	There is a provisional WHO Guideline Value of 0.003 mg/l in drinking water for total cadmium. The MCL of USEPA for cadmium is 0.005 mg/l. The EU Drinking Water Directive sets a standard for cadmium of 0.005 mg/l.
Options for Management in Catchment	The migration of cadmium in the aquatic environment can be controlled by physical, chemical, biological-ecological methods. However, remediation of waters containing cadmium is costly. Cadmium transmission is less under alkaline conditions, thus pH adjustment is an effective and feasible way to reduce cadmium transport in waters.
Options for Treatment at Source	As cadmium(II) is a free cation in aqueous solutions of drinking water, an adsorbent with negative charge, such as MnO_2 or activated carbon, can remove cadmium to a certain extent. In small scale water treatments plant, enhanced coagulation by pre-oxidation using permanganate or ferrate is the most feasible technology. Manganese dioxide formed in-situ plays a main role in removing cadmium when permanganate is employed. Ion exchange can also be used to remove cadmium from drinking water, but this is usually at high cost.
Options for Treatment at Point of Use	At a point of use, cationic ion exchange may be considered as the most appropriate method to remove cadmium. Some technologies, such as nanofiltration and activated carbon, can also be selected to provide safe water at a small scale.

CHROMIUM

Occurrence	Chromium occurs naturally in some sedimentary deposits and volcanic areas. Large quantities of chromium ore can be found in Southern Africa, Kazakhstan, Russia and India. There are many industrial uses for chromium and its salts, including pigments, fungicides, ceramics, glass and photography as well as chromium metal production and metal plating. Contamination of waters used for drinking in industrial areas is not uncommon, and there have been a number of high profile cases around the world, with concentrations many times safe limits.
Chemistry	Chromium is a shiny, hard grey metal that is highly resistant to corrosion. It has a complex chemistry, and can exist in oxidation states ranging from +1 to +6, with +3 and +6 being the most common. Chromium compounds have varying solubility, but those with chromium (VI) are most soluble.
Health and Aesthetic Significance	Food is the main source of exposure to chromium. Toxicological data is rather inconclusive, although there is evidence of carcinogenicity for exposure via inhalation. Chromium (VI) is likely to be the most toxic. Chromium salts do not impart a taste or odour to water.
Regulatory Standards and Guideline Values	Standards are generally set for total chromium, as it can convert between oxidation states both outside and inside the human body. There is a provisional WHO Guideline Value of 0.05 mg/l in drinking water for total chromium. The USEPA sets an MCL for chromium of 0.1 mg/l. The EU Drinking Water Directive sets a standard for chromium of 0.05 mg/l.
Options for Management in Catchment	Plumes of chromium (VI) can spread for considerable distances through contaminated aquifers. Remediation of contamination is possible, and there are a number of remediation schemes underway around the world. Remediation is costly and many sites suffer from historic contamination, giving complicated liability issues.
Options for Treatment at Source	Chromium (III) is easier to remove than chromium (VI) as it is less soluble in water. Therefore traditional large-scale water treatment process are likely to be more effective at removing chromium (III). Ion exchange would appear to be the most reliable option for chromium (VI) removal, although there is little evidence of this being done directly on drinking water supplies and the hexavalent form may best be removed by first converting it to the Cr (III) form or by using a weak base anion resin on a one pass basis. Further treatment technologies are the subject of some study.
Options for Treatment at Point of Use	At a small scale, cationic ion exchange may be the most feasible technology and there are some proprietary processes that claim to do this. As with all treatment technologies it is important that the selected process is approved for use with drinking water and accredited for the desired application.

COBALT

Occurrence	Cobalt is ubiquitous at low quantities in the environment although concentrations vary. Almost never occurring in native form, cobalt is usually combined with elements such as arsenic and sulphur. The main locations where cobalt is mined are Zambia, Congo, Canada, Russia and Australia. Cobalt is primarily used for its magnetic properties, to make alloys with other metals and for jet engine and cutting tool production. Radioactive isotopes of cobalt, which do not occur naturally, are used in medicine. Contamination of environmental water in industrial areas is documented, and has been detected at 426 out of the 1,636 National Priority List waste sites designated by the US EPA. A small proportion of these were radioactive isotopes. In natural waters concentrations up to 36 microgrammes per litre have been documented. Concentrations up to 50 mg/l have been reported in urban and mining areas.
Chemistry	Cobalt is a hard grey metal with magnetic properties. Oxidation states from −1 to +4 are possible, but in the environment +2 or +3 are the most common. In water the +2 form is readily oxidised to +3, from where adsorption and precipitation occurs readily.
Health and Aesthetic Significance	Cobalt is an essential element for human health, being a constituent of vitamin B12. Severe health effects from exposure to naturally occurring cobalt are rare, with most effects reported due to inhalation. Ingestion of large quantities of cobalt may cause effects on the blood and liver as well as dermatitis.
Regulatory Standards and Guideline Values	There is no WHO Guideline Value for cobalt. The USEPA has not set an MCL for cobalt. The EU Drinking Water Directive does not set a standard for cobalt.
Options for Management in Catchment	Cobalt is unlikely to be an issue for drinking water, and where it is detected is likely to be associated with industrial pollution and other contaminants of far greater significance. The location of an alternative source is likely to be the most effective option.
Options for Treatment at Source	Cobalt should be readily removed by most conventional water treatment processes, especially those involving sedimentation or adsorption. Some pre-oxidation to Cobalt's +3 oxidation state may be necessary, but should be easily achieved.
Options for Treatment at Point of Use	In the unlikely event that point of use treatment for cobalt is required, an absorptive process is likely to be the most cost effective option. Successful removal of cobalt using bentonite and diatomaceous earth sorbants has been reported. Cationic ion exchange is also likely to be successful, and some selective resins are available for industrial water treatment.

COPPER

Occurrence	Copper is a red metal that is relatively common in nature. It often occurs in sulphide or carbonate form, sometimes in association with other metals such as iron and arsenic. Copper has many uses in industry, including electronics and as an ingredient in pesticides and fungicides. It is also commonly used in water pipes and fittings. If present in drinking water at significant concentrations, the most likely source of copper is dissolution from pipework and plumbing fittings, and this source should be considered initially. High concentrations of copper in drinking waters as supplied are rare, but concentrations of several milligrammes per litre are achievable within properties where copper pipework is present and conditions are favourable to dissolution. A survey published by Health Canada found that treated water concentrations across the country did not exceed 100 microgrammes per litre, but maximum concentrations up to 560 microgrammes per litre were detected at consumer taps due to the influence of household pipework. Dissolution of copper fittings and pipework is especially likely in waters where either pH is low or pH is high and low in alkalinity.
Chemistry	Copper exists primarily in monovalent and divalent forms, although +3 and +4 oxidation states also exist. In environmental water the prevalent form of copper varies depending on pH. Copper strongly adsorbs to clay particles and interacts with organic material such as humic acids to form complexes.
Health and Aesthetic Significance	Copper is an essential human nutrient. The acute effects of excessive copper are primarily gastrointestinal. Chronic health effects are less well defined. Aesthetically, copper will impart a metallic taste to the water at concentrations. In studies, the taste threshold has been documented as 0.0024–0.0026 mg/l. Blue green staining of water and sanitary ware can occur. Under certain circumstances even hair may take on a blue/green tinge. At higher concentrations, water itself can take on a blue appearance. If any of these are noted, the presence of copper should be strongly suspected and further investigated through sampling.
Regulatory Standards and Guideline Values	USEPA sets an MCLG for copper of 1.3 mg/l but regulates through use of the Lead and Copper Rule to control the corrosivity of the water. Europe sets a maximum concentration of 2 mg/l. WHO has a health-based guideline value of 2 mg/l.
Options for Management in Catchment	Although less likely than corrosion of plumbing, the possibility that copper could be present in the source water should at least be considered. This is especially the case where sources of industrial pollution are likely to be present or in mining areas and where natural copper deposits are known. Unless very localised, it is likely that removal or remediation of industrial and mining contamination at source will be expensive and difficult, and the identification of an alternative source of water is likely to be preferable. Where multiple sources of water exist, each should be sampled to establish whether one is significantly more contaminated and could be abandoned, reducing overall copper concentrations to within acceptable limits.

Options for Treatment at Source	Treatment for copper is likely to be focussed around conditioning water to inhibit corrosion of copper pipework. Theoretically at least, reverse osmosis could be used to reduce copper concentrations in incoming water, however this approach should be a last resort after all catchment options have been considered. Conditioning of water is best achieved where the water enters the property, unless significant amounts of copper pipework exist upstream. The aim is to change the water chemistry to enable a thin carbonate or oxide film to form on pipework, preventing further corrosion. This is usually achieved by increasing the pH and alkalinity of the water using a filter containing alkaline media such as limestone or calcite chips, or by dosing a carefully controlled quantity of sodium carbonate or sodium bicarbonate. This is an effective approach but clearly should only be used where installation and ongoing monitoring and maintenance are likely to be sufficient – a poorly operated dosing installation risks being ineffective at best and overdosing chemicals causing health risks at worst. The dosing of sodium hydroxide solution could also prove effective, however some sources consider the risks associated with handling and controlling dosing of this chemical to be too great for a water supply run without professional expertise.
Options for Treatment at Point of Use	If copper removal at a single tap is required, point of use reverse osmosis may be an option. Where water is hard, some form of pre-treatment may be required to prevent excessive scaling of the membrane. If the majority of copper pick-up is occurring within the property, flushing of taps prior to use may be a simple way of significantly reducing copper concentrations.

IRON

Occurrence	Iron is the most common element on Earth by mass. Many iron minerals have been found on the Earth's surface, including magnetite, hematite, siderite, limonite, pyrite, ilmenite and maghemite. Iron and its alloys are by far the most common metals in industrial use, as well as daily life. Its compounds are also widely used. Dissolved iron in water is derived naturally from soils and rocks. It may also result from the corrosive action of water on unprotected iron or steel pipes and tanks. Ground waters often contain a certain amount dissolved iron all over the world. Surface waters may also occasionally contain appreciable amounts of iron originating from industrial wastes or acid run-off from mining operations.
Chemistry	Elemental iron occurs in meteoroids and other low oxygen environments because iron tends to oxidize in the presence of oxygen and water. It exists in a wide range of oxidation states, −2 to +6, although +2 and +3 are the most common. Fe^{2+}, which is soluble, is found in well waters and in waters with low levels of dissolved oxygen. Iron oxides are common compounds which are widespread in nature and readily prepared in the laboratory. They are present in almost all of the different compartments of the global system: atmosphere, pedosphere, biosphere, hydrosphere, hydrosphere and lithosphere.
Health and Aesthetic Significance	Iron is a necessary trace element in the human body. Iron-containing enzymes and proteins, often containing heme prosthetic groups, participate in many biological oxidations and in transport, such as hemoglobin, cytochrome, and catalase. Iron deficiency can cause anemia, as well as damage of psychological activity and intellectual development. However, large amounts of ingested iron can cause iron poisoning. Excess iron typically damages cells in the heart, liver and elsewhere. Iron in water supplies causes aesthetic and operational problems, such as bad tastes and colour, staining and deposition in the water distribution system leading to high turbidity. On small supplies using ultra-violet (UV) disinfection it can deposit on the surface of the UV lamp casing, greatly reducing effectiveness and potentially putting health at risk.
Regulatory Standards and Guideline Values	There is a provisional WHO Guideline Value of 0.3 mg/l in drinking water for total iron. The MCL of USEPA for iron is 0.3 mg/l. The EU Drinking Water Directive sets a standard for iron of 0.2 mg/l.
Options for Management in Catchment	Iron reaches ground water by rain, surface and waste water filtration dissolving minerals from soil strata. Iron may also be present as a result of the use of iron coagulants or the corrosion of steel and cast iron pipes during ground water extraction and distribution. Water percolating through soil and rock dissolves iron, and these minerals subsequently enter groundwater supplies.

In anaerobic conditions, such as in ground water deposits, iron is reduced to its soluble oxidation state Fe^{2+} that can be easily dispersed but it is oxidized to the insoluble state Fe^{3+} in aerobic conditions. Surface water does not usually contain high concentrations of iron because the oxygen-rich water enables iron minerals to settle out as sediments. The ubiquitous nature of environmental iron makes it difficult to manage iron in catchments, although the avoidance of acidic conditions, such as stagnant water or mine drainage, helps to keep the iron in insoluble form and out of the aquatic environment.

Options for Treatment at Source

There are a variety of methods which can remove iron including filtration, ion-exchange, membrane processes, adsorption, biological treatment, and bio-filtration. Ferrous iron oxidation followed by solid-liquid separation is the most commonly used method for iron removal. Since the solubility of ferric iron in solution is very small, ferrous iron is firstly oxidized to ferric iron by oxygen, permanganate, ferrate, or ozone and the formed ferric iron can be separated from water using solid-liquid separation including sedimentation and/or filtration. Many treatment suppliers provide relatively simple loose media filters which can remove insoluble iron and are backwashable on a timer system.

Options for Treatment at Point of Use

Many small-scale methods for iron removal are available, using filtration, cationic ion-exchange, membrane processes, chemical oxidation, and adsorption, biological treatment, and biofiltration and membrane filtration. Where iron concentrations are relatively low, simple and cheap filtration with replaceable cartridge filters may be possible.

LEAD

Occurrence	Lead may occur naturally in water supplies or it may be the result of man-made pollution. A further major source is the dissolution of lead from pipework and plumbing fittings. In older properties this can be by far the commonest reason for lead to be detected in drinking water samples. Solders and brass plumbing fittings can contain surprisingly high concentrations of lead which is available to be leached into the water. Sources of industrial pollution include lead smelting, heavy metal processing, paints, batteries, and in some countries, fuels. Naturally occurring lead exists mainly as compounds with sulphur and oxygen as well as carbonates. Galena, lead sulphide, is the main ore. Principle deposits of lead are Australia, USA, China and Canada.
Chemistry	Lead is a soft, dense, silvery metal which rapidly tarnishes to a dull grey on exposure to the air. It is the heaviest stable element, with an atomic number 82. On exposure to water it can react to form protective coatings, the solubility of which depends on water chemistry. Where conditions favour this, significant quantities of lead can dissolve into drinking water. Temperature and pH affect this tendency significantly in addition to water chemistry.
Health and Aesthetic Significance	Lead has no known use in the human body. It is a cumulative general poison, with both acute and chronic effects. Children under 6 and pregnant women are most at risk. Health effects can be wide ranging, but the impact on neurodevelopment in children is especially well documented. No safe lower threshold for lead concentration has been identified. Consequently, WHO guideline values have progressively reduced and may potentially be lowered further in future years. The current guideline value is 10 microgrammes per litre. Lead may impart a sweet taste to waters.
Options for Management in Catchment	Lead is rarely present in raw waters, although this is not impossible (refer to case study in Section 9.4). Elevated lead concentrations may occur where lead bearing minerals are present. Lead is often associated with mineral veins, so source waters may not have a uniform distribution of the metal and it may be possible to exclude a particularly lead-rich source. In such circumstances, given the absence of lower threshold for adverse health effects, it may be most appropriate to abandon the source. Where point source contamination is present, such as due to localised waste disposal, this may be more easily dealt with.
Options for Treatment at Source	Treatment at source is unlikely to be appropriate for the reasons described above. Management of lead in water supplies is mostly centred around removal of lead pipework and fittings, remediation of point source contamination in the catchment. If treatment is considered, coagulation and filtration processes, ion exchange, reverse osmosis or adsorptive technologies may all be successful.
Options for Treatment at Point of Use	Elevated lead concentrations at the point of use are most likely to be due to the presence of lead in plumbing. Small water supplies are especially vulnerable to the use of non-approved plumbing materials and techniques. This cause should always be explored in the first instance, even in circumstances where lead pipework is not thought to be present, as it is by far the simplest to resolve. As a last resort, where no clear source of lead is identified, a GAC filter may be used to remove the lead ions.

MANGANESE

Occurrence	Manganese is unevenly distributed on the Earth. It makes up about 1000 ppm (0.1%) of the Earth's crust, making it the 12th most abundant element. It is often found in combination with iron, and in many minerals. More than 95% of manganese mines exist in The Republic of South Africa, Russia, Gabon, Australia, Brazil, and India. The common minerals of manganese are pyrolusite, psilomelane, hausmannite, braunite, rhodochrosite and manganblende. Manganese and its compounds are widely used in various production areas. About 90% to 95% of manganese is used in the steel industry, and 5% to 10% of manganese is used in other areas such as the chemical industry, light industry, building materials industry, defence industry and electronics industry.
Chemistry	Metallic manganese is a hard, crisp, active, and silvery white metal. Manganese exists in several different states ranging from −3 to +7, and the common states in nature are +2, +3, +4, +6, and +7. The most stable oxidation state for manganese is +2 and many manganese(II) compounds are known, such as manganese(II) sulfate ($MnSO_4$) and manganese(II) chloride ($MnCl_2$). The +3 oxidation state is known in compounds like manganese(III) acetate. Mn(IV) is insoluble and exists as brown or black precipitation in neutral solutions. Manganate(VI) can be produced by dissolving Mn compounds in molten alkali while exposed to air. Permanganate (Mn(VII)) is a commonly used laboratory and water treatment reagent because of its oxidizing properties.
Health and Aesthetic Significance	Manganese is an important metal for human health, being absolutely necessary for development, metabolism, and the antioxidant system. Nevertheless, excessive exposure or intake may lead to a condition known as manganism, a neurodegenerative disorder that causes dopaminergic neuronal death and parkinsonian-like symptoms. Some recent research has linked elevated manganese concentrations with developmental disorders of the nervous system. Manganese precipitation gives water a brown-black colour when exposed to air, and can make water aesthetically unacceptable. Manganese gives water a metallic taste.
Regulatory Standards and Guideline Values	The Chinese national drinking water standard for Mn(II) is 0.10 mg/l. WHO recommends an upper limit of 0.05 mg/l for manganese in water for drinking. The European Union and United States Environment Protection Agency (USEPA) set regulatory upper limits of 0.05 mg/l.
Options for Management in Catchment	In an oxidative environment, manganese tends to form insoluble precipitations and its concentration in the relevant environment is sufficiently low. While in a reducing environment, manganese tends to be reduced to soluble Mn(II), which can spread for considerable distances through contaminated aquifers. Lots of methods are proposed for the remediation of manganese contamination including physical, chemical and biological methods. Though remediation can be complex and costly, phytoremediation appears to be a promising method because it is comparatively cheap and efficient.

(*Continued*)

MANGANESE (*Continued*)

Options for Treatment at Source	There are a series of treatment processes aimed at manganese removal such as oxidation, filtration, bio-filtration and adsorption, etc. Among these processes, oxidation including chemical oxidation and biological oxidation is a classical technology which can be used in the industrial scale water treatment engineering and scaled for smaller applications. Aeration alone is unlikely to be sufficient to achieve complete oxidation of any soluble manganese. Chemical oxidation employing chlorine, ferrate or permanganate may be economic and convenient for small scale water treatment plants. Bio-filtration is a widely used method in large scale plants for manganese removal. Alternatively, enhanced coagulation is emerging as a promising process for the removal of manganese by using permanganate or ferrate peroxidation in Chinese surface waters.
Options for Treatment at Point of Use	Ion exchange and nanofiltration would appear to be the most reliable solution for manganese treatment at the point of use, though they are costly compared with other methods. Efforts are being made to develop low-cost technologies for small scale requirements.

MERCURY

Occurrence	High concentrations of naturally occurring mercury are rare, but do occur and have been documented in the USA and Japan. Occurrence will commonly be with other metals or in volcanic areas. Mercury may also be associated with coal deposits. Contamination of water from industrial activities is possible – mercury is used in electrical products, especially batteries and fluorescent light bulbs, including those commonly used to generate UV light to disinfect small water supplies. Mercury deposits of commercial significance have been found in Spain, USA, Kyrgyzstan, Italy and Slovenia. The most significant producer or mercury is now China.
Chemistry	Mercury is a metal, liquid at ambient temperatures. It occurs as both organic and inorganic compounds, with some natural processes creating methyl mercury. Inorganic mercury is commonly found in two oxidation states, I and II, with II being the most common state in nature. Mercury(II) chloride is readily soluble, whereas mercury(I) chloride and mercury sulphide being much less so.
Health and Aesthetic Significance	Mercury is a well-known toxin. Health effects are primarily neurological and in the kidney. Acute exposure to high concentrations can cause sever damage to the heart and stomach. Organic forms of mercury are more toxic to humans and difficult to remove from drinking waters.
Regulatory Standards and Guideline Values	WHO has set a guideline value for inorganic mercury of 0.006 mg/l. USEPA has an MCL of 0.002 mg/l. The EU Drinking Water Directive sets a limit of 0.001 mg/l.
Options for Management in Catchment	Where mercury is likely to be naturally occurring, such as in volcanic areas, near metalliferous mineral deposits, or coal measures, it may be advisable to seek an alternative water source could be sought. Unless widespread industrial pollution is suspected, man-made contamination could well be from point sources, such as buried batteries or electrical equipment and this option is probably the most productive one to be investigated initially.
Options for Treatment at Source	If the source of the mercury cannot be located, GAC filters or ion exchange technologies are likely to be the most cost-effective methods for removal of mercury at source. GAC is most likely to be effective for organic forms of mercury. Mercury selective ion exchange resins are commercially available. Reverse osmosis membranes should also be effective. There is some evidence that lime softening may be effective, by precipitating the mercury as a hydroxide.
Options for Treatment at Point of Use	Point of use treatment is likely to be feasible, based on the technologies above. GAC filtration may be simplest, however a service contract or close monitoring to ensure timely media replacement would be advisable. The spent media may contain high levels of contaminant and need special disposal measures. Point of use cationic ion exchange and RO units are also available.

NICKEL

Occurrence	Nickel occurs widely in the environment as it has high mobility. There are a number of naturally occurring ores, primarily oxides and sulphides, including chalcopyrite and pyrrhotite. It often occurs with other metals such as iron and manganese. Nickel is not usually found in native form in nature. Man-made sources of nickel are mainly focussed on smelting, refining and steel production. It is often used to make alloys with other metals and in plating. Australia and new Caledonia have the largest ore deposits. Kettles, and nickel or chromium plated plumbing fittings may all contribute significant quantities if nickel into water.
Chemistry	Nickel is a silvery metal with atomic number 28. It has a number of oxidation states and forms compounds with all the common anions.
Health and Aesthetic Significance	Nickel is not thought to be acutely toxic, but chronic exposure may cause heart and liver damage. Some individuals may become sensitised to nickel, causing dermatitis.
Regulatory Standards and Guideline Values	Who Guideline Value is 0.03 mg/l. USEPA sets an MCl of 0.1 mg/l for nickel. The EU Drinking Water Directive sets a standard of 0.2 mg/l.
Options for Management in Catchment	When faced with a detection of nickel in a water supply, unless obvious catchment sources are present, it is recommended to firstly consider the possibility of artificial contamination from metallic pumps and fittings in the supply system. Similarly, it may be possible to locate a point source of nickel in the catchment such as waste metal. If this can be removed it may save expensive treatment costs.
Options for Treatment at Source	Chemical precipitation may be effective at removing nickel, especially at high pH. Selective ion exchange resins are available, although unlikely to be a cheap option for the small scale user. Various adsorbents are known to be effective – GAC may be sufficient, or specialist metal oxides which are available on the market may give a higher rate of removal.
Options for Treatment at Point of Use	For point of use applications, adsorption is likely to be the most cost effective and low maintenance option, although quality will need to be monitored for breakthrough through the filter.

SELENIUM

Occurrence	Selenium is a relatively uncommon trace element, often found associated with other metal ores. It has a number of industrial uses, primarily in electronics and photography. It has also sometimes been added to fertilisers and can accumulate in soils in this way. Industrial pollution can cause selenium to be present from mining wastes, metal refining and combustion products of coal and petroleum. Parts of the US and Canada have significant concentrations of naturally occurring selenium. Significant concentrations of selenium in drinking water have also been documented from China.
Chemistry	Selenium is a metalloid substance with atomic number 34. It has a strong association with sulphur. Elemental selenium and selenium sulphide are relatively insoluble and therefore thought to be less of an issue in drinking water, however sodium salts of selenium are soluble and may be present. It is likely to be most problematic in alkaline situations, where it is oxides to soluble selenite form.
Health and Aesthetic Significance	At low concentrations selenium is an essential element, however it does exhibit toxic properties when present in larger quantities. Main health effects comprise long term circulatory damage resulting in effects at the periphery of the body, such as hair and fingernail loss and impairments to the peripheral nervous system. A wide variety of other impacts have also been noted, including respiratory, cardiovascular and gastrointestinal effects. Aesthetic impacts of high selenium concentrations in drinking water are likely to be limited, although some selenium compounds have a characteristic garlic odour and the odour has been reported on the breath of individuals who have consumed a significant quantity of selenium.
Regulatory Standards and Guideline Values	WHO has set a guideline value for selenium of 0.01 mg/l. This has been replicated in many national standards including the EU Drinking Water Directive. The USEPA sets a higher Maximum Contaminant Level Goal for selenium of 0.05 mg/l, although a number of individual states adopt the lower WHO value.
Options for Management in Catchment	Catchment management options for selenium are likely to be limited. Where it is naturally occurring, it is likely to be fairly ubiquitous, although alternative water sources could be sought. Where selenium is not known to occur naturally, industrial pollution should be suspected, especially if there is evidence of mining, smelting or manufacturing activities in the catchment. Remediation of contaminated land may be possible, but it is likely to be significantly cheaper and easier to abandon the source.

(*Continued*)

SELENIUM *(Continued)*

Options for Treatment at Source	Selenium in its oxidised Se(VI) form, such as selenite, is most difficult to remove. Other forms are likely to be relatively easily removed from drinking water. WHO reports that adsorptive technologies using iron or aluminium oxides are successful. Ion exchange technologies have also been used with good results, as has reverse osmosis. Base anion exchange systems have been shown to perform effectively on a large scale and it is likely that this is one of the most practical treatment solutions for small water supplies. Due to the chemical similarities between selenium and sulphur, it is suggested that sulphate will also be removed by the resin, making the process less economic for high sulphate waters. The USEPA guidance on selenium suggests that coagulation and filtration processes may be effective for selenium removal, although other sources contradict this. It is likely that the efficacy of this process depends strongly on the chemical form of the selenium. Coagulation is unlikely to be an appropriate solution for most small supplies in any case.
Options for Treatment at Point of Use	Point of use treatment is likely to be feasible, based on the technologies above. Filtration with activated alumina or iron oxide are likely to be simplest as they require minimal maintenance and no chemical regeneration, however a service contract or close monitoring to ensure timely media replacement would be advisable. Point of use anionic ion exchange and RO units are also available.

TIN

Occurrence	Tin is a silvery white metal. Its principal ore is cassiterite. Most industrial uses are in plating, especially food containers. It is also used to make alloys with other metals, and can be a component of solder. Organotin biocides are used, especially as anti-fouling agents in the marine industry. Compared to ingestion from food, especially canned food, water does not represent a significant route of human exposure. Tin ores have been mined in a number of locations, including Cornwall, Bolivia, South East Asia and Australia.
Chemistry	Tin exists primarily in oxidation states +2 and +4. Most tin compounds are insoluble, limiting concentrations in drinking water. Tin halides are the most common salts.
Health and Aesthetic Significance	Tin has limited effect on the human body and is poorly absorbed. At high concentrations, gastric irritation is possible, but this is unlikely to occur via drinking water.
Regulatory Standards and Guideline Values	WHO does not consider tin in drinking water to be a significant issue and has not set a guideline value. Most countries have not set a standard for tin in drinking water.
Options for Management in Catchment	Unlikely to be an issue, primarily due to its low solubility. Organic tin compounds could occur where historic use as antifouling agents has taken place, and it is likely that seeking an alternative source of water would be the best approach, especially of the source of contamination is diffuse.
Options for Treatment at Source	Treatment for tin is unlikely to be required, but ion exchange would probably be the most cost effective solution.
Options for Treatment at Point of Use	Point of use cationic ion exchange is the most viable technology.

TUNGSTEN

Occurrence	Tungsten is a grey/white metal with exceptional strength and high melting point. In the environment it is found as a compound with other elements, its most common ores being wolframite and scheelite. China is the largest producer of tungsten ore, with quantities also mined in Canada and Russia. It is unlikely to be encountered frequently in drinking water. Where it does occur naturally, deposits are likely to be localised. Due to its strength tungsten has many uses in alloys with other metals. Tungsten's very low solubility in most forms meant that it has not historically been considered a large risk in drinking water, however there is some suggestion that this assumption is not consistently correct, and that under certain conditions aqueous forms of tungsten may be an issue. Significant concentrations of tungsten in drinking water wells have been found in parts of Nevada.
Chemistry	Oxidation state +6 is the most common form, although all other states above –2 are possible.
Health and Aesthetic Significance	Tungsten's health impacts due to ingestion have not been fully quantified, and what information exists is primarily related to occupational exposure. There is the suspicion that oral ingestion of tungsten may cause developmental abnormalities in foetuses. It may also accumulate in bones.
Regulatory Standards and Guideline Values	WHO does not set a guideline value for tungsten. Most countries have not set a standard for tungsten in drinking water.
Options for Management in Catchment	Most likely man-made sources of contamination are existing or former metal working plants, although tungsten's use as a metal alloy means almost any industrial area working with metallic substances could be a source. Firing ranges and war zones (tungsten is used in bullets) could also result in environmental contamination.
Options for Treatment at Source	Little has been established, but USEPA suggests that traditional coagulation and clarification processes are very effective at lower pH (<6). Of course, these technologies are not usually suited to the smallest supplies. Using a water soluble polymer to complex anionic forms of tungsten prior to removal by ultrafiltration has also shown some success.
Options for Treatment at Point of Use	Unknown – cationic ion exchange methods may be worth exploring once the chemistry of the tungsten present has been understood.

URANIUM

Occurrence	Uranium is a radioactive silver metal. It is widespread in small quantities in nature, usually being associated with granite geology. The most common ore is uranite, formerly known as pitchblende, which is an oxide of uranium. It is often found with tin, copper and arsenic minerals. Canada, Australia and Kazakhstan are the largest producers of uranium. Uranium's main use is as fuel in nuclear reactors. A number of studies have found uranium in drinking water, mainly from private water supplies, at concentrations greater than 20 µg/l, with concentrations up to 700 microgrammes per litre being detected at one source in Canada.
Chemistry	Uranium occurs in oxidation states ranging from +2 to +6, with the latter being by far the most common. Many ores of uranium are largely insoluble in drinking water, although carbonate based minerals are soluble and these are likely to provide the most likely source of natural concentrations of uranium in drinking water.
Health and Aesthetic Significance	The primary concern with elevated concentrations of uranium in drinking water is its toxicity rather than any radiological effects. Uranium is known to have an acute effect on kidney function. Wider hazards associated with uranium in drinking water relate to its decay to radon, a significantly radioactive gas which presents a risk to health via inhalation. Good ventilation in areas where uranium is known to occur in drinking water is important.
Regulatory Standards and Guideline Values	WHO does sets a provisional guideline value for uranium of 30 microgrammes per litre, although it acknowledges that there is a lack of firm evidence on which to base a numerical standard. This value supersedes an earlier, lower, guideline value of 15 microgrammes per litre. The USEPA sets an MCL standard of 30 microgrammes per litre for uranium. The EU currently does not set a standard for uranium in drinking water, although this may be under review.
Options for Management in Catchment	Where uranium occurs naturally it is likely to be ubiquitous across the area and an alternative type of source should be sought. Point source contamination due to industry such as the combustion of coal products could be remediated or an alternative, less contaminated source sought.
Options for Treatment at Source	Although removal by coagulation and sedimentation methods may be possible, studies suggest it is highly pH dependent. It is also less likely to be practical for small water supplies. As uranium is only a risk when ingested (it cannot be inhaled or absorbed through the skin), point of use treatment at taps used for drinking may be a more effective control measure. The risk from radon gas should also be considered in any supply where uranium is present in significant concentrations.
Options for Treatment at Point of Use	Point of use treatment options for uranium consist mainly of reverse osmosis and anionic ion exchange. The former may be simpler for domestic situations as minimal chemical consumables are required, although neither options are maintenance free. Adsorption onto GAC or other media such as a form of titanium oxide may be effective where uranium concentrations are lower and could be considered for its simplicity, although removal rates are less certain than for the other methods.

VANADIUM

Occurrence	Vanadium is ubiquitous across the natural environment at trace concentrations. Higher concentrations can occur with certain geological conditions, including volcanism and hydrothermal activity. Higher concentrations of naturally occurring vanadium may be found in Hawaii and parts of Italy. Industrial uses of vanadium are in the manufacturing of steel, batteries, ceramics and magnets. These industries may provide a localised source of contamination.
Chemistry	Pure vanadium is a bright white, ductile metal. It occurs in association with other elements rather than in native form. Vanadium is relatively insoluble in water. It can occur in several oxidation states, with +5 being the most common, although states +4, +3 and +2 also occurring.
Health and Aesthetic Significance	Primary effects of ingestion of high concentrations of vanadium include nausea and stomach cramps. There is some evidence that vanadium is a possible carcinogen.
Standards	WHO does not set a guideline value for vanadium The USEPA has not set an MCL standard for vanadium, although this may be reviewed. The EU currently does not set a standard for vanadium.
Options for Management in Catchment	Control is primarily around locating a water source with a lower vanadium concentration, although this may not be possible in certain geological conditions.
Options for Treatment at Source	Vanadium is not usually considered a problem in drinking water, consequently information on treatment technologies is sparse. Cationic ion exchange processes have been shown to reduce vanadium concentrations. Some proprietary adsorptive media is also available to treat vanadium in drinking water.
Options for Treatment at Point of Use	Small-scale versions of the above technologies may be available for point of use treatment.

ZINC

Occurrence	Zinc is a bluish-white metal in its native form. In the environment it most commonly occurs naturally as a sulphide or carbonate mineral, with sphalerite (zinc sulphide) being the most common. Zinc is often associated with deposits of other metals such as lead, copper and silver. Ore deposits are quite widely distributed around the world, including China, Australia, Peru, Europe and Canada. Zinc has many industrial uses, especially as alloys with other metals to inhibit corrosion – a process known as galvanisation. One of the commonest ways for zinc to enter small water supply systems is via dissolution from plumbing fittings or pipes. Many fittings contain brass and galvanised steel water pipes are not uncommon – these will readily leach zinc, particularly in very soft, low pH waters.
Chemistry	Zinc has an atomic number 30 and an ozidation state of +2. Zinc is a relatively reactive metal and rarely found in its native state. The solubility of zinc compounds varies – sulphides and carbonates (the commonest ores) are not soluble, however zinc chloride, sulphate and nitrate readily dissolve.
Health and Aesthetic Significance	Zinc is an essential element for humans, and most health issues are focussed on a deficiency of zinc rather than an excess. Adverse effects of an excess of zinc are centred around gastro-intestinal issues. An excess of zinc in drinking water – at concentrations greater than 3 mg/l[3] can be detrimental to the appearance of the water, with a surface film developing and an unpleasant metallic taste.
Regulatory Standards and Guideline Values	WHO does not set a guideline value for zinc, whilst acknowledging that higher concentrations may be aesthetically undesirable. The USEPA sets a secondary maximum contaminant level (SMCL) for zinc of 5.0 mg/l. Canadian provinces have also set a 5.0 mg/l standard for zinc. The EU Drinking Water Directive does not set a standard for zinc.
Options for Management in Catchment	Where the source of zinc is man-made, the source should be identified and removed if possible. Where naturally occurring zinc is present it is likely that an alternative source will be required, following investigation.
Options for Treatment at Source	The most likely source of zinc in small water supplies is dissolution from pipes and fittings. Where water is corrosive it can be conditioned prior to entering the distribution pipework by raising the pH and mineral content. Where possible, old galvanised pipework should be replaced with plastic alternatives approved for use with drinking water. It is unlikely that it will be practical or necessary to remove zinc from a water supply, and efforts should focus on prevention of the issue. Precipitation using lime softening or ion exchange resins are most likely to be the most practical options. Reverse osmosis membranes, while expensive, could also be effective.
Options for Treatment at Point of Use	At the tap, cationic ion exchange filters are most likely to be practical, once sources from pipes and fittings have been eliminated.

REFERENCES

Aluminium

Water Quality Association (2010). Aluminium Fact Sheet. http://www.wqa.org/Portals/0/Technical/Technical%20Fact%20Sheets/2014_Aluminum.pdf (accessed June 2015).

Arsenic

World Health Organisation (2012). Media Centre Fact Sheet – Arsenic. Fact sheet no. 372 http://www.who.int/mediacentre/factsheets/fs372/en/ (accessed June 2015).

United States Environmental Protection Agency. Basic Information About the Arsenic Rule. http://water.epa.gov/lawsregs/rulesregs/sdwa/arsenic/basic-information.cfm (accessed June 2015).

Smedley P. L. and Kinniburgh D. G. (2002). A review of the source and behaviour of arsenic in natural waters. *Applied Geochemistry* **17**(5), 517–568.

Ahmed F. (2001). An overview of arsenic removal technologies in Bangladesh and India. In: Technologies for Arsenic Removal from Drinking Water. Bangladesh University of Engineering and Technology; United Nations University, Dhaka, pp. 251–269

Beryllium

United States Environmental Protection Agency. Basic Information About Beryllium. http://www.who.int/water_sanitation_health/dwq/chemicals/BERYLLIUMsummarystatement.pdf?ua=1 (accessed June 2015).

New Hampshire Department of Environmental Services (2007). Beryllium in Drinking Water. http://des.nh.gov/organization/commissioner/pip/factsheets/dwgb/documents/dwgb-3–3.pdf (accessed June 2015).

Cobalt

Ministry of Environment, British Columbia, Canada (1981). Ambient Water Quality Guidelines for Cobalt (http://www.env.gov.bc.ca/wat/wq/BCguidelines/cobalt/cobalt_over.html (accessed June 2015).

Copper

World Health Organisation (2004). Copper in Drinking-water Background Document for Development of WHO Guidelines for Drinking-Water Quality. WHO, Geneva, Switzerland.

Health Canada (1992). Environmental and Workplace Health – Copper. http://www.hc-sc.gc.ca/ewh-semt/pubs/water-eau/copper-cuivre/index-eng.php#fnb10 (accessed June 2015).

US NRC (2000). Copper in Drinking Water. Washington, DC, National Research Council, National Academy Press.

Dvorak B., Skipton S., Woldt W. and Drda S. (2013). Nebguide – Copper. University of Nebraska. http://www.ianrpubs.unl.edu/pages/publicationD.jsp?publicationId=789 (accessed June 2015).

Lead

World Health Organisation (2011). Lead in Drinking-Water Background Document for Development of WHO Guidelines for Drinking-Water Quality. WHO, Geneva, Switzerland.

Mercury

World Health Organisation (2005). Mercury in Drinking-Water Background Document for Development of WHO Guidelines for Drinking-Water Quality. WHO, Geneva, Switzerland.

Selenium

World Health Organisation (2011). Selenium in Drinking-Water Background Document for Development of WHO Guidelines for Drinking-Water Quality. WHO, Geneva, Switzerland.

Yang G., Yin S., Zhou R., Gu L., Yan B., Liu Y. and Liu Y. (1989). Studies of safe maximal daily dietary Se-intake in a seleniferous area in China. II. Relation between Se-intake and the manifestation of clinical signs and certain biochemical alterations in blood and urine (published erratum appears in Jour. Trace Elem. Electroly. Health Dis 1989. 3:250). *Jour Trace Elem. Electroly. Health Dis.* **3**, 123–130.

British Columbia Government, Water Land and Air Protection (2001). Water Quality Guidelines for Selenium. http://www.env.gov.bc.ca/wat/wq/BCguidelines/selenium/report2c_v3–06.htm#TopOfPage (accessed June 2015).

Smith K., Lau A. and Vance F. Evaluation of Treatment Techniques for Selenium Removal. Degremont Technologies Report IWC 09–05. http://www.degremont-technologies.com/IMG/pdf/tech_infilco_Selenium-Removal.pdf (accessed June 2015).

Tungsten

Walker M. and Fosbury D. (2009). Arsenic, As((III)), and Tungsten in Nevada County's Private Water Supplies. *J Water Health.* 2009 Jun; **7**(2), 293–301.

United States Environmental Protection Agency (2014). Technical Fact Sheet – Tungsten. http://www2.epa.gov/sites/production/files/2014–03/documents/ffrrofactsheet_contaminant_tungsten_january2014_final.pdf (accessed June 2015).

Uranium

World Health Organisation (2004). Uranium in Drinking-Water Background Document for Development of WHO Guidelines for Drinking-Water Quality. WHO, Geneva, Switzerland.

Health Canada (2010). Water Talk – Uranium. http://www.hc-sc.gc.ca/ewh-semt/pubs/water-eau/uranium-eng.php (accessed June 2015).

Zinc

Australian Government, Department of the Environment. National Pollutant Inventory – Zinc and compounds. http://www.npi.gov.au/resource/zinc-and-compounds (accessed June 2015).

Mayock D., Fawell J., Merrington G. and Watts C. (2008). Review of England and Wales Monitoring Data for Which a National or International Standard Has Been Set. Defra Project Code: CEER 0703 DWI 70/2/215 WT1207. DWI, London, England.

World Health Organisation (1996). Zinc in Drinking-Water Background Document for Development of WHO Guidelines for Drinking-Water Quality. WHO, Geneva, Switzerland.

Chapter 9

Case studies

9.1 ARSENIC REMOVAL IN SMALL WATER SUPPLIES IN ITALY

Sabrina Sorlini, Michela Biasibetti and Francesca Gialdini

9.1.1 Plant A: Adsorption with granular ferric hydroxide

Plant A is located in Central Italy. The water source is contaminated by arsenic (40 μg/l) that could be essentially ascribed to natural phenomena, even though some human sources are still under examination. For this reason, a drinking water treatment plant (DWTP) for arsenic removal was built and operated since 2006. This plant treats spring water and well water (about 50 m deep) corresponding to an average flow rate of about 2 l/s. The project flow rate is much higher (6 l/s), allowing the plant to treat the maximum flow rate during the spring season. Figure 9.1 shows Plant A and Table 9.1 shows the average water quality parameters in raw water and in treated water.

Table 9.1 Plant A: Water quality.

Parameter	Unit	Raw Water	Treated Water
pH	–	7.4	7.2
Conductivity at 25°C	μS/cm	495	530
As	μg/l	40	<0.2
Fe	μg/l	65	3
Mn	μg/l	7	9
SO_4^-	mg/l	25	24
Cl^-	mg/l	34	–

Figure 9.1 Plant A: GFH filters.

The plant applies the process of adsorption on Granular Ferric Hydroxide (GFH) by means of 2 filters operating in series (Figure 9.2): the first one is a coarse-filter and the second one a polishing-filter. The filtration system could operate both in sequence A-B and in the reverse sequence B-A to maximize the substitution/regeneration time of GFH: the polishing-filter turns into a coarse-filter when the latter is exhausted and refilled by virgin/regenerated material. Treated water at the outlet of the plant is mixed with another groundwater source that is not treated, but only disinfected with sodium hypochlorite.

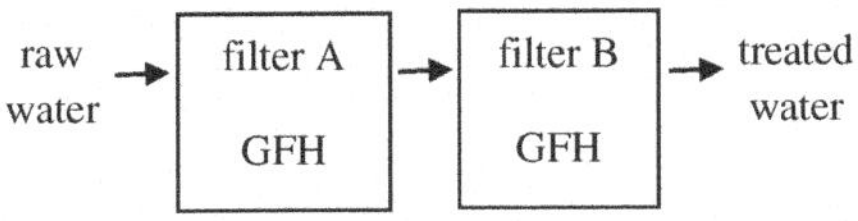

Figure 9.2 Schematic diagram of plant A.

From the bottom to the top, each filter bed is composed of:

- collecting system consisting of a nozzle plate;
- sand draining-bed;
- GFH adsorbing-bed (grain size 0.3–2 mm).

Each filter has a 15 m^3 volume and a 2.8 m diameter. Filters need backwashing once/twice a week to move the filter-bed and remove suspended solids eventually retained in the filter media. Moving the filter-bed is important to prevent "short-circuits" that could occur over time in the filter-bed, leading to a significant reduction in the arsenic removal efficiency. Backwashing is carried out by using pre-filtered water, collected in the final storage tank. The ratio between backwashing-water and processed-water is lower than 1%.

The plant does not present specific phases for residue treatment; moreover the filter backwashing water is directly discharged into the public sewers. As a matter of fact, Regional Regulations in Tuscany (L.R. 31–5–2006, no. 20 and D.P.G.R. 8–9–2008, no. 46/R) do not identify backwashing-water as waste. Nonetheless, the quality of this residual flux is consistent with the quality required for the release in surface water bodies as stated by Legislative Decree 152/2006, Tab. 3.

The efficiency of arsenic removal is monitored by a specific sampling program of the inlet and the outlet of the DWTP (Table 9.2). The duration of the adsorbent material is about 54 months.

Table 9.2 Efficiency of arsenic removal.

Arsenic in Raw Water (2010 Data) [μg/l]	Treated Volume [$m^3 \times 1000$]	Treated Volume [BV]	Arsenic Adsorbed [g/kg]	Arsenic in Treated Water [μg/l]
40.9	277	19	0.68	<0.2

On the basis of the analytical results, considering the mass of GFH present in the filters and the water flow rate, this DWTP treated about 18,000 BVs of water and reached a removal capacity equal to 0.68g/kg.

Regarding the costs related to plant A, the only data available is the total cost of the plant, which is equal to €120,000. The main Operational and Maintenance (O&M) cost for GFH filtration plants is substitution/regeneration of the filter-bed. Electric power cost, related to power consumption by backwashing pumps, is almost negligible. Moreover this kind of plant operates in automatic mode by using a PLC, so labour cost for regulation/control activities (generally once a week for each plant) is very low (Collivignarelli *et al.* 2011).

9.1.2 Plant B: Resins functionalized with ferric hydroxide

Plant B is located in Central Italy and treats groundwater contaminated by arsenic concentration variable from 20 to 40 μg/l. The plant, treating about 2l/s flow, has a simple treatment chain: a pumping station, two adsorbent resins functionalized with ferric hydroxide (FeOOH) operating in series and a disinfection phase with NaClO (Figure 9.4). Figure 9.3 shows Plant B and Table 9.3 shows the average water quality parameters in raw water and in treated water.

Figure 9.3 Plant B: Filter with functionalized resin (left) and final disinfection (right).

Table 9.3 Plant B: Water quality.

Parameter	Unit	Raw Water	Treated Water
pH	–	7.2–7.3	7.2–7.3
As	μg/l	20–40	<1
Fe	μg/l	11	12
Mn	μg/l	40	3
Silica	μg/l	25	25

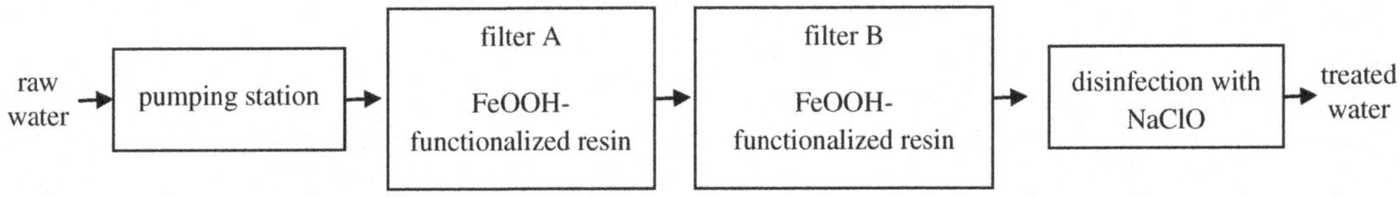

Figure 9.4 Schematic diagram of plant B.

Filter A is a coarse-filter and filter B a polishing-filter: the media of both the two filters is composed of polymeric resins functionalized with granular ferric oxides thus combining the processes of ion exchange and adsorption. This allows to extend the life of the material if the water is rich in interfering ions (above all chlorides and sulphates) and to remove trivalent arsenic. Phosphates, vanadium, chromium and silica can interfere with this material, but these compounds are not present in the water.

Inside the 1.5 m filter column with a 0.8 m diameter, the resin layer is about 1 m high. The resin has a complete adsorption capacity on As(V), but a lower removal capacity on As(III).

Filters need backwashing at least once a month, according to the total suspended solid concentration in the inlet water, using raw or treated water. The discharge is made directly into a surface water body, after storage for the suspended solids sedimentation. The backwashing is semiautomatic, made by the operator command, according to the target yield of removal for suspended solids. The annual backwashing water volume is about 44 m^3.

The resins are regenerated on average once a year, according to the arsenic concentration values registered at the outlet of the first filter. After the resin regeneration, the volume of the spent brine is estimated to be approximately 12 m^3. This plant does not present specific treatment for the residue generated from water purification, but a treatment for backwashing water mixed with the exhausted brine could be implemented in the future. The efficiency of arsenic removal is monitored by a specific sampling program of the inlet and the outlet of the DWTP: Table 9.4 reports arsenic concentration in raw and treated water during filter operation with a water flow of 170 m^3/d at 500 Bed Volume. According to these results, the technology reached an arsenic removal capacity higher than 96% (Collivignarelli *et al.* 2011).

Table 9.4 Efficiency of arsenic removal.

Arsenic in Raw Water [µg/l]	Arsenic at the Outlet of the Second Resin [µg/l]	Arsenic in Treated Water [µg/l]
40	<5	<1

Regarding the costs related to the water treatment line, the initial investment accounts for 2 cent€/m^3 of treated water, reagents account for 1.5 cent€/m^3 of treated water, energy for 0.24 cent€/m^3 of treated water, personnel for 0.05 cent€/m^3 of treated water and maintenance for 0.05 cent€/m^3 of treated water. The total cost is about 5.84 cent€/m^3 of treated water.

9.1.3 Plant C: Chemical precipitation and adsorption with granular ferric hydroxide

Plant C, located in Northern Italy, has been operating since 2007 treating groundwater (about 200 m deep) with an average flow of about 2l/s, even if the plant can treat a maximum water flow of 5.5l/s. The average inlet arsenic concentration is about 37 µg/l. The plant is composed of the following treatments: aeration, biological filtration, oxidation with H_2O_2, precipitation with $FeCl_3$, sand filtration and, for the 50% of the inlet flow, adsorption on synthetic Granular Ferric Hydroxide (GFH); the remaining 50% of the flow is by-passed and mixed with the flow at the outlet of the GFH treatment stage and it is sent to the final disinfection stage with chlorine dioxide (Figure 9.6). Figure 9.5 shows Plant C and Table 9.5 shows the average water quality parameters in raw water and in treated water.

Figure 9.5 Plant C: Sand filtration (left) and GFH filtration (right).

Table 9.5 Plant C: Water quality.

Parameter	Unit	Raw Water	Treated Water
pH	–	7.8	7.7
Conductivity at 20°C	μS/cm	397	399
As (III)	μg/l	37	3
As (V)	μg/l	7	–
Fe	μg/l	96.1	28
Mn	μg/l	67.3	<5
NH_4^+	mg/l	1.7	<0.05
nitrates	mg/l	0.3	–
SiO_2	mg/l	19	–

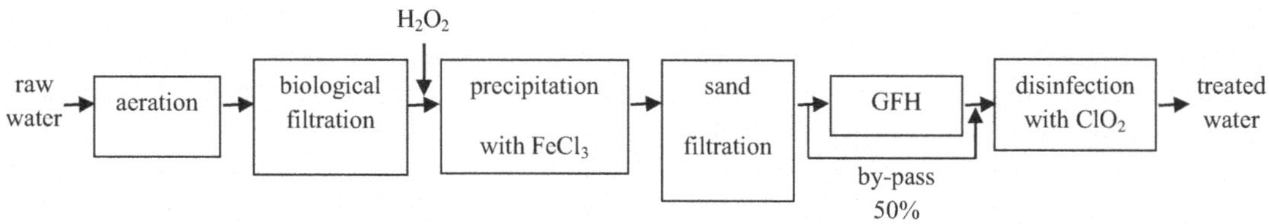

Figure 9.6 Schematic diagram of plant C.

The first treatment stage is the oxygenation by injection of compressed air in a storage tank. The oxygen necessary for the biologic process is at least 500g(O_2)/h, supplied by a 5.5 kW compressor that guarantees an air flow of 150 Nm^3/h with a maximum pressure of 5bar. The biological filtration oxidizes ammonia and manganese using a biofilter with a 2,500 mm diameter, a 2,000 mm high, a sand bed 1.5 × 2.0 mm, 1,600 mm high.

The subsequent oxidation operates with 4–5 L(H_2O_2)/h in order to maintain a redox potential equal to 400–500 mV, obtaining an arsenic removal lower than 10 μg(As)/L.The sand filtration removes the insoluble flocs containing arsenic, iron and manganese using a multilayer filter (silica sand of different grain, hydroantracite and pyrolusite) with a media with the following geometrical dimensions: 2,600 mm diameter, 2,500 mm high and 1900 mm depth.

The GFH filter (2,100 mm diameter, 2,500 mm high, 1,000 mm adsorbent material depth) treats the 50% of the flow, because the filter is able to remove As concentration till 1 μg/L, therefore the mixing of the two flows allows to respect the legislative limit, while maintaining a long useful life of the adsorbent material. Since the treated water flow is low and the amount of substances to be removed is minimum, in this case the filter backwashing once a month is enough.

The plant does not present a specific line for the residue treatment, but there is only a storage tank for the collection of the backwashing water generated from the biological filter, the sand filter and the GFH filter.

The arsenic removal yield in plant C is about 92%. Regarding the costs related to the water treatment line, reagents account for €0.028/m^3 of treated water, energy for €0.149/m^3 of treated water, manpower for €0.087/m^3 of treated water, mechanical parts spare for €0.062/m^3 of treated water, substitution/reintegration of filtering materials for €0.156/m^3 of treated water. The total cost is about €0.482/m^3 of treated water (Collivignarelli *et al.* 2011).

REFERENCES

Collivignarelli C., Riganti V. and Sorlini S. (2011). L'arsenico nelle acque destinate al consumo umano. Esperienze e applicazioni delle tecnologie di rimozione dell'arsenico e aspetti gestionali (Arsenic in drinking water. Experiences and applications of arsenic removal technologies and management aspects). Dario Flaccovio Ed.

Sorlini S. and Collivignarelli C. (2011). Arsenic in water for human consumption. Technical solutions for arsenic removal from drinking water. LAP LAMBERT Academic Publishing.

9.2 A NEW BOREHOLE SUPPLY WITH IRON REMOVAL FOR A SINGLE PROPERTY IN ENGLAND, UK

John Murphy, Springhead Water Consulting, Oldham and UK

9.2.1 Background

The case study outlines the process of installing a new water supply for a single property on a large estate in Derbyshire. It was believed the property, a small cottage, had been fed by a surface water supply since it was first built in 1855. Investigations found a map, a sketch and notes of the system dated 1954 and showed the supply as a spring supply and that this fed into a collection tank approx 1.0 km from the property. This pipe was shown as galvanised iron pipe in 1954 with some sections of lead pipe within a cast iron duct. The pipe was now however believed to be mainly 3/4" or 1" black alkathene pipe laid approx 4 feet below surface level. The spring collection tank, close to the spring, was on a slightly higher level than the roof tank in the cottage. This minimal gravity head coupled with the long pipe run resulted in a very low gravity feed head to the property. A risk assessment had shown the spring would be an unsafe supply from a microbiological aspect and considerable source protection would be required to improve the safety of the supply. There would also be a need to run a new larger diameter feed pipe from the source to the property and although the use of a storage tank and booster pump at the cottage could have increased the pressure there was some uncertainty that the supply would be adequate to meet the requirements of a modern holiday let, which was the proposed new use of the cottage. The cottage had been used as a single residency for a number of years with a very low water demand. The notes relating to the supply also showed it was prone to blocking, presumably from sediment, and required back flushing on a regular basis.

Mains water could be provided but this would mean installing a new dedicated water booster pump station and installing a new private mains pipe to feed the supply a height of approx 100 metres over distance of approximately 600 metres and would require a break tank and booster pump to be located adjacent to the mains tapping and then a further storage tank and booster pump at the cottage. As a result of these investigations into the requirements to improve the existing supply or provide an alternative supply, it was decided to drill a bore hole and use that as the prime water source and retain the existing surface supply as a means of feeding the animal feed troughs on the pasture close to the cottage and retain the ability to use the spring supply for agricultural use if required.

The area close to the cottage was agricultural pasture land use for gazing livestock. There was also a wooded area close to the cottage where pheasant and grouse were plentiful as well as deer from the estate. The property was situated close to the highest point of the local area at 224 metres above sea level. Approximately 1.4 kilometres away at a slightly higher level than the cottage was a small lake which was locally believed to be fed by artesian water. There was also a bore hole drilled within the previous twelve months to feed another property on the estate 3.4 kilometres from the proposed bore. The depth of this existing bore was 60 metres and this had shown iron levels of 4090 μg/l and manganese levels of 676 μg/l in the raw water. Although the initial level of both these metals had been lower when the bore was first drilled, the water had been successfully treated with filtration at the property. The local geology showed that the area for the proposed bore was largely sandstone, as had been the case on the nearby bore supply. A local well driller with experience in the area was commissioned to drill the bore and it was proposed that provision be made during the drilling process in case the bore was artesian.

9.2.2 Network requirements

The cottage was to be converted into a two bedroom luxury dwelling. Heating would be supplied by a ground heat recovery system. It was decided not to drill the bores for ground water heat recovery but to use a below ground pipe matrix recovery system. One of the reasons for this was the unknown effect of the water recharge of a ground water heat recovery system and if this would have a detrimental effect of the proposed bore which would be providing the drinking water. The cottage was also to have a new septic tank system installed and, in line with recommendations, was sited down slope from the bore at a suitable distance.

The new bore supply was to supply all the domestic water needs for the cottage. It was anticipated that the supply would probably contain some iron and manganese based on the results of the nearest bore and therefore may require a treatment plant.

The use of a power shower within the property would result in large peak water flow requirements. In order to keep treatment costs as low as possible, and reduce energy requirements, it was planned to feed the water into a storage tank at the property with any equipment for iron or manganese removal being installed prior to this tank. The bore would feed into the treatment filter and then the storage tank at a flow considerably lower than the required peak flow rates and thus any filter required to treat the iron and manganese would be smaller than that required to deal directly with the high peak flow to the property. It was planned to fit a sediment filter and ultra violet disinfection unit downstream of the tank and for this to treat all the water to the cottage. This would have to be capable of treating the peak flow requirements of the system. The booster pump required to feed water from the storage tank through the sediment filter and UV and pressurise the whole of the system was a small variable speed unit which would give a constant pressure of approx 3 bar to the system and alter speed and hence energy requirements as flow varied. This ensured energy requirements for the booster pump would be optimised over the large variations in water flow requirements for the cottage.

The booster pump and UV unit had to be designed to meet the peak flows to the property of both hot and cold water as the system was a pressurised system. The peak flow was estimated at 70 litres per minute for short periods, 10 to 15 minutes, mainly due to the power shower requirements. The cottage would provide accommodation for up to four people. Experience View of the Cottage had indicated that on luxury holiday sites 500 litres/person/day water usage was often the required demand. A higher than normal usage was reinforced by data on the British Water web site showing flow and loadings for small rural sewage systems which detailed considerable variation in flow for varying grades of establishments with a figure of 300 litres/person/day for prestige establishments. It was felt that this provided additional data based on an assumption of "what comes out is about the same as what goes in." This is volume is considerably in excess of the quantities advised in the Private Water Regulations of 200 litres/person/day.

Part of an existing outbuilding adjacent to the property was to be used as a location for the storage tank and treatment equipment as this minimised the planning changes that would be required. This in practice placed a maximum size on the storage tank that could be installed of 1100 litres.

9.2.3 Drilling the borehole

The data in Table 9.6 is extracted from the daily drilling log kept by the well driller. The first 12 metres was drilled with a 12 1/4" bit. This was then lined with a casing which was grouted into position. From this depth drilling continued using 8" drill. Test pumping at 91 metres showed the flow from the bore was not substantial and it was decided to continue drilling. The drill log shows "loads" of water was present when the depth of 121 metres was reached. The bore was fitted with screens. A pump was lowered using 1" galvanised steel pipe to a depth of 118 metres and a flow rate of 900 litres per hour was achieved. The bore was completed and capped (Figure 9.7). The pump was run for a number of hours each day over the period of approx 7 days and did not run dry during any of these periods.

Table 9.6 Plant C: Water quality.

Depth From	To	Comments
Ground Level	0.5 metres	Top soil
0.5 metres to	9.0 metres	Sandstone
9.0 metres	10.0 metres	Red sandstone
10.0 metres	11.80 metres	
11.80 metres	16 metres	Grey sandstone
16 metres	25 metres	Yellow sandstone bands
25 metres	29 metres	Yellow and orange sandstone
29 metres	34 metres	Red sandstone more shale
34 metres	63 metres	Shale
63 metres	65 metres	Bits of everything, shale and sandstone all colours also first signs of water
65 metres	91.47 metres	Shale; well dipped and water level at rest was 77 metres from ground level
91 metres	121.47 metres	Shale. Drilling was recommenced to obtain more water. At 121 metres drilling stopped and lots of water was present

Figure 9.7 Well head showing bore and cap with pump electrical isolation switch.

Under normal circumstances the flow tests would have carried on for longer. However to meet the schedule required to complete the work and the target dates for opening the holiday let it was decided to carry out an initial water sample from the bore at this point. This was to highlight any potential problems and determine if iron and manganese was a problem. Field test kits had shown elevated iron levels and a slightly acid pH. The water appeared very clear with no noticeable odour. In some cases hydrogen sulphide is also a problem on bores but this supply seemed free from this problem.

9.2.4 Time line

- 16th April 2012 Start of drilling on site
- 20th April 2012 Completion of drilling and installation of pump
- 23rd April 2012 Run flow tests on bore and start to draw water off on a daily basis
- 1st May 2012 Carry out sample tests on untreated bore water supply. These tests showed high
- levels of iron (823 μg/l) in the water and some microbiological failures.
- May 2012 Chlorination of bore carried out by well driller.
- Water used on a regular basis for ongoing non potable use on site
- Installation of water storage tank, booster pump, sediment filter and ultra
- violet unit. Connection of bore supply to tank, treated water to cottage and initial
- running of the system
- 28th May 2012 Re-test bore, post UV, and tap at cottage supplies. Iron levels on bore
- had increased dramatically to 29,500 μg/l but all the parameters on treated supply
- were satisfactory.
- June 2012 Install iron reduction filter on the supply and commission.
- July 2012 Cottage completed and available as holiday let
- Dec 2012 Risk assessment on system completed

- Dec 2012 Service on equipment and change UV lamp
- June 2013 Interim service on equipment
- July 2013 Annual analysis for 2013 carried out
- Nov 2013 First service carried out on iron reduction filter
- Dec 2013 Service on pump and UV, lamp changed on UV

9.2.5 Analysis

The analysis data for the bore supply is shown in (Table 9.7a); the initial sample drawn on 1st May 2012 gave high levels of iron, 823 µg/l., above the maximum allowable drinking water limit of 200 µg/l. This was not unexpected as the nearest bore to this one had also shown high iron levels. The pH of 6.47 was also not unusual for this area.

The sample showed some microbiological failures and, and considering the bore was only recently completed, that was not entirely unexpected. The well driller was asked to chlorinate the bore as a precaution. In other aspects the analysis gave satisfactory data and it was decided to carry on running the bore on a daily basis whilst further work in upgrading the property was carried out. The water storage tank was installed and a water booster pump, sediment filter and ultra violet unit were installed downstream of the new storage tank. These items were included on the original specification and would be required as part of any proposed specification of treatment plant for the system.

The bore was pumped for periods throughout the May and a further analysis taken on 28th May 2012. By this time the water storage tank had been installed together with the booster pump, sediment filter and UV unit. It was decided to take samples of the raw water from the bore before it entered the storage tank, of water immediately after the UV, and at the kitchen tap in the property for further analysis

The water drawn from the bore was initially very clear with no smell or taste detected. There was however some noticeable iron staining in the storage tank from the bore supply and the water within the tank was cloudy. The water sample from the bore started to discolour to an orange colour within approximately one hour of being drawn despite the fact that the sample bottle had as much air as possible excluded from the it.

The samples from after the UV and from the tap appeared very clear and did not exhibit any discolouration. A separate sample taken after the UV showed a transmission efficiency of 98% at 254 nanometres wavelength, the bactericidal wavelength of UV units, when compared to that of deionised water. The iron level of the bore was found to be incredibly high at 29,500 µg/l. This was reanalysed by the laboratory to verify the high reading. This was the highest that I had ever come across on a borehole supply but I was aware that a colleague had experienced figure higher than this on a supply, although that was over 40 kilometres away. This high iron level represented a serious challenge any proposed treatment plant. It was noted that the sediment filter was at this stage reducing the iron level to within parameter requirements but it was heavily loaded with iron from the water after just a few weeks operation and the tank, as mentioned, was already showing heavy staining from the iron.

There were no microbiological or other parameter failures post UV treatment and the bore raw water sample was now also clear from *Clostridium perfringens*, coliforms, *E. coli* and *Enterococci* although there was a high 3 day colony count on the bore water.

An iron reduction filter was selected and installed on the system. The filter was an aerating unit with a specific iron reduction media as the filter bed. Filter details are detailed later in this case study. Checks on the post filter iron levels showed the filtered water to be below 10 µg/l before entering the storage tank using field test kits. The storage tank was cleaned and flushed to remove any iron staining from the initial usage.

Table 9.7a Initial water quality.

Parameter	Regulatory Standard	Bore Raw Water	Bore Raw Water	Kitchen Tap	UV Outlet
		1st May 2012	**28th May 2012**	**28th May 2012**	**28th May 2012**
Aluminium	200 μg/l	155	–	2.1	–
Ammonium	0.50 mg/l	<0.015	–	0.03	–
Antimony	5 μg/l as Sb	<0.01	–	<0.01	–
Arsenic	10 μg/l as As	<0.18	–	<0.18	–
Boron	1000 μg/l as B	18	–	28	–
Cadmium	5 μg/l as Cd	0.007	–	0.01	–
Chloride	250 mg/l as Cl	12.7	–	13.4	–
Chromium	50 μg/l as Cr	1.4	–	1.2	–
Clostridia	nil/100 ml	3	–	0	0
Colour	20 mg/l Pt/Co	0.41	–	2.11	–
Conductivity	2500 μS/cm@20C	192	–	301	–
Copper	2000 μg/l as Cu	<0.7	–	26	–
Cyanide	50 μg/l	1.5	–	<0.3	–
E.coli	nil/100 ml	11	0	0	0
Enterococci	nil/100 ml	3	0	0	0
Fluoride	1.5 mg/l as F	0.1	–	0.16	–
Hydrogen ion (pH)	6.5 to 9.5	6.47	6.84	7.07	7.38
Iron	200 μg/l as Fe	823	29500	73.8	165
Lead	25 μg/l as Pb(10 μg/l from 25th Dec 2013)	0.2	–	21	–
Manganese	50 μg/l as Mn	16	–	42.3	–
Nickel	20 μg/l as Ni	1.8	–	2.7	–
Nitrate	50 mg/l as NO_3	19.2	–	17.5	–
Nitrite	0.5 mg/l as NO_2	<0.009	–	0.158	–
Odour	Acceptable/no abnormal change	0	–	5	–
Sodium	200 mg/l as Na	8.3	–	11	–
Sulphate	250 mg/l as SO_4	25.3	–	27.1	–
Total Coliforms	nil/100 ml	24	0	0	0
Total Oxidised Nitrogen	–	19.2	–	17.7	–
Turbidity	4 NTU	1.75	2.59	2.8	1.96
TVC 3 day 22°C	no abnormal change	532	>1000	11	0
Non lactose fermenting B	–	40	0	0	0

Throughout the first year of operation periodic checks were made on the iron levels using field test kits. All these showed compliance with the regulatory requirements with levels substantially below the 200 μg/l value. There were no signs of discolouration within the storage tank or on the sanitary ware within the property. It was noted on one visit that a small inflatable pool had been brought by one visitor to the cottage and filled indicating the precaution of using higher than normal water demands in assessing the water usage.

Table 9.7b Further water quality data.

Parameter	Regulatory Standard	Bore Raw Water	Kitchen Tap
		November 2013	**July 2013**
Aluminium	200 μg/l	3.7	–
Ammonium	0.50 mg/l	<0.015	–
Antimony	5 μg/l as Sb	–	–
Arsenic	10 μg/l as As	–	–
Boron	1000 μg/l as B	–	–
Cadmium	5 μg/l as Cd	–	–
Chloride	250 mg/l as Cl	–	–
Chromium	50 μg/l as Cr	–	–
Clostridia	nil/100 ml	0	–
Colour	20 mg/l Pt/Co	–	–
Conductivity	2500 μS/cm@20C	–	–
Copper	2000 μg/l as Cu	–	–
Cyanide	50 μg/l	–	–
E.coli	nil/100 ml	0	–
Enterococci	nil/100 ml	0	–
Fluoride	1.5 mg/l as F	–	–
Hydrogen ion (pH)	6.5 to 9.5	7.7	7.25
Iron	200 μg/l as Fe	<3.6	2200
Lead	25 μg/l as Pb(10 μg/l from 25th Dec 2013)	0.15	–
Manganese	50 μg/l as Mn	<2.6	42
Nickel	20 μg/l as Ni	–	–
Nitrate as NO_3	50 mg/l as NO_3	10.9	–
Nitrite	0.5 mg/l as NO_2	0.012	–
Odour	Acceptable/no abnormal change	0	–
Sodium	200 mg/l as Na	11.6	–
Sulphate	250 mg/l as SO_4	–	–
Total Coliforms	nil/100 ml	–	–
Total Oxidised Nitrogen	–	10.9	–
Turbidity	4 NTU	<0.7	–
TVC 2 day 37°C	no abnormal change	0	–
TVC 3 day 22°C	no abnormal change	2	–

The analysis of the treated water supply in 2013 (Table 9.7b) showed the supply was again compliant with the parameters tested including the microbiological requirements. Iron level on the treated water was shown to be below 3.7 µg/l.

9.2.6 Equipment data

Borehole pump

The borehole pump was an electrically driven submersible centrifugal pump and was lowered into the bore on 1" galvanised pipe to a depth of 118 metres

Iron reduction filter

The filter selected was a Shakesby ED range filter designed to operate on float switch control in the storage tank (Figure 9.8). The filter is a fully aerating unit using air entrainment via venturi built into the control head. This filter is capable of dealing with flow rates of up to 1.3 m^3/hour and has a backwash flow requirement of 0.6 m^3/hour, within the capability of the supply from the bore.

Figure 9.8 Storage tank and iron reduction filter installed on the system.

The venture produces an air flow in service of approximately 8.5 litres/minute, and 8.0 litres/minute during backwash. The air entrainment in the filter allows the iron to oxidise out of solution within the filter vessel and this is of prime importance in the filtration operation. This oxidation of anaerobic water was what was observed when the samples were drawn from the bore on 28th May and the water took on the orange appearance.

The air entrainment on the backwash results in a very vigorous backwashing of the filter media at flow rates below the service flow rates. A shut off valve within the filter head prevents backwash and rinse water entering the storage tank during this sequence. The filter media used is a specific iron reduction media provided with the filter by the manufacturer.

Tank

The storage tank was selected to fit into the outbuilding and provide capacity to mitigate the peak flow demands on the system. It is a WRAS approved tank of 1100 litre capacity.

Booster pump

The booster pump was selected to supply flows of up to 80 litres/minute at 3 bar pressure. In order reduce energy consumption the unit selected had an inverter motor drive which allowed the motor to vary its speed over the wide range of flows required for the property whilst maintaining a constant pressure in the network. The pump also had dry run protection which would automatically shut the pump down to prevent it being damaged in the event of the storage tank running dry.

Sediment filter

A sediment filter with a 5 micron cartridge filter was installed after the pump and before the ultraviolet unit to prevent any sediment passing into the network. This was also a UK Water Regulation Advisory Service (WRAS) approved item.

Ultra violet unit

A WRAS approved UV unit was selected for this system in line with Regulation 5 of the 2009 Private Water Regulations for England. The unit was selected based on the flow requirements for the property and the requirements to provide and energy level of 40 mJ/cm^2 at the end of lamp life. A transmission test on the water was also carried out which showed the transmissivity to be in excess of 98% for the water. A flow restrictor was installed downstream of the UV to ensure the design peak flows were not exceeded.

9.2.7 System flow tests

On completion of the tank, pump and filter installation flow tests were carried out to measure flow rates down stream of the UV unit and restrictor. This confirmed the maximum design flow rates could not be exceeded. Tests carried out in the property showed that good flow and pressures were achieved at all outlets.

9.2.8 Risk assessment

A risk assessment of the completed system was carried out in December of 2012. As a result of this minor changes were made to provide better labelling of valves and include a schematic diagram of the system within the treatment room.

9.2.9 Review

On most private supplies the resources compared to a mains water utility are much more constrained. These are often constraints in terms of source option and treatment methods. There may also be limitations in maintenance and monitoring resources requiring a different approach to meeting challenges.

In this case the decision to rely on a borehole supply presented some significant challenges. It was anticipated that there could be both iron and manganese levels that would require treatment. The initial bore tests looked promising in that although the iron levels exceeded the drinking water parameters the level measured was treatable and there was no high manganese levels. Experience has suggested that a bore supply make take a considerable period to settle and provide a consistent water quality. This can present problems in selecting equipment due to variations in water quality and in some cases adequacy. Decisions on the specification of any iron reduction filter were delayed as long as possible with the bore operating as much as practical before a final specification was achieved. The iron level of the bore at 29.5 mg/l found on the second analysis on 28th May was unexpected and very high. The water was very clear as it discharged from the bore and only suffered from the iron precipitation after coming into contact with oxygen in the surrounding air, indicating anaerobic water with the iron totally in solution. The rapid orange discolouration also gave some encouragement that that the water could be treated by aeration and filtration.

After discussions with the filter manufacturer a fully aerating unit was sized and installed. The water supply has operated without any water quality issues since it was installed. The 2013 audit analysis showed very low iron levels of below 3.6ug/l on the treated supply.

- A check on the raw bore supply when the first filter service was carried out in November of 2013 showed an iron level of 2200 μg/l and a manganese level of 42 μg/l. The raw water iron level was dramatically lower than that measured at the end of May 2012. The tests on the filtered water showed an iron level of 30 μg/l, a manganese level of 9 μg/l and a pH of 7.0
- By comparison the nearest bore which was also on the Estate and was checked on the same day in 2013 showed an increase in raw water iron levels from 4070 μg/l to 5900 μg/l, and an increase in manganese levels from 678 μg/l to 1070 μg/l.
- Mechanically the equipment operated well. The pump had automatically shut down in a few instances. This was believed to be a problem associated with the pressure control settings and was corrected by increasing the capacity of the pump pressure vessel from 20 litres to 60 litres.
- It was decided to change the UV lamp in December 2012 rather than leave for the full 12 months in order to bring the lamp change into line with all the other UV units on the various private supplies on the estate. These are all changed in December of each year to reduce the possibility of wrong sequencing of the change.
- Intermediate services of the equipment such as checking pump operation, sediment filter changes along with pipe and system integrity checks are carried out at six month intervals
- Experience on small supplies such as this indicate that caution should be placed on the initial analysis a monitoring of both treated and raw water is of value. Whilst this is routine on large public supplies there is no requirement on private supplies to monitor the raw water quality of a supply.

In this case the routine monitoring of both raw and filtered water for iron, manganese and pH levels is carried out during the servicing of equipment using field test kits and the water supply at the property is tested annually in line the regulatory requirements with samples being analysed by an accredited laboratory.

ACKNOWLEDGEMENTS

The Estate Office, Trustees of Chatsworth Settlement
PR Marriott Drilling
Shakesby Filters
Marshall Pump Systems

9.3 METALS IN SMALL WATER SUPPLIES IN AREAS OF WATER SCARCITY IN AFRICAN REGIONS

Gea Oliveri Conti, Letizia Ferlito†, Alfina Grasso*, Giovanni Arena*, Chiara Copat* and Margherita Ferrante**

*LIAA. Laboratory of Environmental and Food Hygiene-Department "G.F. Ingrassia" – Hygiene and Public Health, Catania University, Italy.
†Consultant for SNV Netherlands Development Organisation.

Water scarcity, defined as the difficulty of obtaining sources of fresh water for use during a period of time, is the lack of sufficient available water resources to meet the demands of water usage within a geographic region (EC, 2014). It already affects every continent and this is a very big problem globally, in fact, more than 1.2 billion people lack access to clean drinking water. As of 2006, Sub-Saharan Africa had the largest number of water-stressed countries of any other place on the world and 300 million, of the 800 million people living in Africa, live in a water stressed environment (WHO, 2007).

Small water supplies continue to be essential to the growth of regional and local economic hubs. Since rural growth centres are of key strategic importance for economic and social development, piped water schemes have been implemented by several governments over the past 20 years and have gathered pace in the current century. This trend has increased rapidly since the early 2000s. In 2010, 15% to 20% of the rural population in African countries lives within the service area of a small piped water or water wells scheme. In countries such as Senegal, Rwanda and Mauritania, it is more than 30%. The sustainable management of hygiene and total quality of these small piped water schemes is therefore becoming increasingly important for African countries growth (Hoang Gia *et al.* 2010).

However, when small water supplies are available in African areas, the poor chemical and microbiological quality of their waters does not always discourage their use, moreover, in most African countries, responsibility for water supply has been transferred by the central government to local governments or to village peoples (Hoang Gia *et al.* 2010).

A survey carried out by our research group in Guinea Bissau – West Africa (the fourth poorest country in the world) provides useful information on the potential impact of human consumption and associated conditions and it is part of the wider and still much debated topic of access to safe drinking water in developing countries and built suitable sanitary conditions (Loo, 2012).

The field survey was carried out in two regions Oio and Cacheu in the north of Guinea Bissau (Figure 9.1) with the aim of providing data on the metals and metalloids occurrence.

In February 2013, 20 samples of drinking water taken from the existing wells and 20 sediment samples, were collected and preserved in situ for a secure transportation at the Environmental and Food Hygiene Laboratories of Catania University. For the extraction procedure, sediments were digested in a microwave system EthosTC (Milestone) with HCl and HNO_3. Hg analysis both in groundwater and sediment was conducted by a FIAS (Flow Injection Analyse System) (Perkin Elmer), Al, Fe, B, Cu, Zn, Ba by an ICP-OES Optima 2000DV (Perkin Elmer), whereas Mn, As, Se, V, Sb, Cd, Cr, Pb, Ni, Co, Be and Mo were detected by an ICP-MS Elan DRC-e (Perkin Elmer) (Copat, 2013).

Fe, Pb, Mn and Al were found occasionally with values greater than the international thresholds for drinking water (Table 9.1) set by WHO and EC (WHO, 2011). In the sediments, our results were under the limits set by the Sediments Quality Guidelines of EPA for certain metals (McDonald, 1994).

The results obtained highlight that there is a close correlation between the chemistry of the water and that of the sediment. Indeed, in different sediment samples detected concentrations of Fe and Al were higher than other metals as well as in some samples of water, due to the nature of the soil of the area, which is mainly composed of Bauxite (AlO) and Laterite (FeO), as already documented in literature.

However, the presence of high metals concentrations in some samples may be mainly connected to the maintenance procedures and to the systems characteristics.

From a socio-economic point of view, both the close correlation between the chemistry of the sediment and the high presence of metals in some drinking water samples are problems due to the lack of adequate facilities for pumping and distribution systems, but also due to the poor maintenance of existing wells that affects the dissolution of alloys used for the construction or extraction.

Our study has shown:

- the important role of maintenance procedures and of construction plan of water pipeline system characteristics,
- potential toxicity issues for the local populations.

Because of its public health impact, drinking water quality is a collective concern and its management is everyone's duty to ensure the safeguard and protection of the water supply through new and old technologies. All those involved, i.e. legislators, controllers and individual users, have an important role in preventing water contamination through correct water use and proper planning of solid waste disposal and civil and industrial sewage drainage. A combination of all mentioned factors would ensure the most effective approach and result in a health gain for the whole of society.

Some important issues, like the identification and assessment of the risks related to chronic, low-level environmental exposure to metals and more complex compounds, are still outstanding and this lack is very problematic in countries with severe water scarcity.

REFERENCES

Copat C., Arena G., Fiore M., Ledda C., Fallico R., Sciacca S. and Ferrante M. (2013). Heavy metals concentrations in fish and shellfish from eastern Mediterranean Sea: consumption advisories. *Food Chem Toxicol.* **53**, 33–37.

European Commission (EC). Environment-Water. 03/11/ 2014. http://ec.europa.eu/environment/water/quantity/about.htm (accessed June 2015).

Hoang Gia L., Fugelsnes T. and Hawkins P. Sustainable Management of Small Water Supply Systems in Africa. Practitioners' Workshop Report, Maputo, Mozambique. October 6–8, 2010. Pp. 1–16. Water and Sanitation Program (Africa).

Loo S. L., Fane A. G., Krantz W. B. and Lim T. T. (2012). Emergency water supply: A review of potential technologies and selection criteria. *Water Res.* **46**(10), 3125–3151.

MacDonald D. D. (1994). Approach to the assessment of sediment quality in Florida coastal waters. vol. 1 – Development and evaluation of sediment quality assessment guidelines. Florida Department of Environmental Protection Office of Water Policy.

WHO. The International Decade for Action: Water for Life 2005–2015. Coping with Water Scarcity World Water Day 2007. http://www.who.int/water_sanitation_health/wwd7_water_scarcity_final_rev_1.pdf.

WHO (2011). Guidelines for drinking-water quality, fourth edition. World Health Organisation Editor. Geneva, Switzerland.

9.4 UNEXPLAINED LEAD CONTAMINATION OF A SMALL WATER SUPPLY IN NORTHERN SCOTLAND

Patricia Sheldon, Environmental Health Department, Highland Council, Scotland and UK

9.4.1 Background

Between 2011 and 2013, the Environmental Health Service of Highland Council, Scotland, dealt with an interesting case relating to lead contamination of a private water supply. Under The Private Water Supplies (Scotland) Regulations 2006, the Local Authority has a duty to risk assess and routinely monitor Type A (larger and commercial) supplies and a rural estate cottage with a deer larder and kennels was identified as meeting this criteria.

The risk assessment for this supply identified a spring fed supply with a large collection chamber approximately 7 metres depth, constructed of cast concrete rings. The chamber was situated in low lying ground, which was not grazed by animals and was surrounded by heather/moss and grassy vegetation. From the chamber, water was pumped to a large plastic storage tank within the grounds of the estate cottage, and thereafter fed under gravity through a central ultra-violet treatment system and 10 μm pre-filter. Testing of the water at the deer larder revealed a lead level of 50 μgPb/l, which at that time was twice the regulatory prescribed concentration value (PVC) of 25 μgPb/l. The current PCV for lead in Scotland is 10 μg/l.

The Environmental Health Officer's (EHO) initial discussions with the Estate were to check that the water distribution system did not contain any lead piping or lead soldering, as this was well documented to cause non-compliance with the lead standard. The Estate excavated most of the pipework at the property and identified alkathene pipes. The fixtures and fittings with the source collection chamber were stainless steel. The Estate advised that the spring supply was constructed approximately 14 years earlier and according to their knowledge, the potablity test undertaken at the time did not identify any concerns about lead (the Regulatory PCV for lead pre-2007 was a PCV of 50 μgPb/l).

Further testing of the supply was carried out by the EHO from various locations on the supply. The lead levels varied from 41 μgPb/l to 73 μgPb/l; all exceeding the PCV at the time of 25 μgPb/l. At this stage of the investigation, the EHOs discussed the possibility that the spring source maybe contaminated.

From Dec 2013, a stricter PCV lead level of 10 μgPb/l was to be implemented in accordance with the *EU Drinking Water Directive,* so a solution to bring the lead levels under this PCV level was required.

9.4.2 Identifying the source of the problem

A search of the Highland Council's contaminated land register of potentially contaminated sites and GIS information did not identify any historical industrial use of the ground. The bedrock in the area was of undifferentiated metamorphic from Grampian Group and Dava Succession and may have naturally higher than normal concentrations of lead, but this supply was sourced from a spring supply, so likely contamination from bedrock was low.

The Estate decided to connect a temporary new spring supply into the existing source chamber; this was captured approximately 1000 metres to the east. The original inlet pipe to the chamber was capped. A visit by the EHO identified that the temporary new spring supply inlet pipe was not flowing so there may have been an obstruction in the pipe. The water level in the chamber was therefore lower than expected.

However, this enabled the officer to observe a second inlet pipe below the water level. A water sample was taken from the source chamber and reported a lead level of 123.8 μgPb/l.

Further investigation by the Estate identified that the supply was constructed with two different spring collection chambers (Figure 9.9) and with the assistance of the Estate plumber, the first collection chamber was located, about 6 metres away from the other; covered by dense vegetation. The ground around this chamber was boggy and prior to excavation, two water samples were taken from surface water puddles. The results were reported as 165.8 μgPl and 211.4 μg/Pl.

Figure 9.9 First spring chamber (left: discovered late into the investigation) and second spring chamber (right).

The newly discovered first spring collection chamber was constructed of cast concrete with a capacity of approximately 2000 litres. Water was pumped out of the chamber to reveal an alkathene inlet pipe which had a steady flow rate of water. The inlet water was sampled and reported at 2454.5 μgPb/l.

This latter sample was conclusive that the source of the lead contamination was from the spring source as oppose to any post collection contamination. The outlet pipe from the first collection chamber was capped and the second spring chamber was left for a few weeks to be fed by the temporary new spring supply prior to testing. A concentration of 31.8 μgPb/l was reported. Whilst the lead level had reduced, it would not comply with the proposed PVC of 10 μgPb/l.

The second spring chamber was pumped out and this identified for the first time that the base of the of the chamber was laid with gravel so there was natural percolation of water into the chamber from the surrounding water table as well as being fed by the temporary spring supply. The continued contamination of the water with lead in the chamber was therefore highly likely.

On discussion with the Estate, it was thought that refuse from the historical occupation of the cottage and former estate/farming operations may have been dumped on the ground, potentially including waste metal or lead-acid batteries, although no records existed to prove this theory. The Estate thereafter disconnected the spring supply and commissioned a new borehole supply, which was sited on higher ground behind the property.

IWA Publishing's authorised EU representative for General Product Safety Regulations is Diane D'Arras, 15 rue Duret, 75116 Paris, France, e-mail: safety@iwap.co.uk.

Printed and bound by CPI Group (UK) Ltd, Croydon, CR0 4YY
13/05/2026
02109604-0012